BEHAVIOR AND MORPHOLOGY IN THE GLANDULOCAUDINE FISHES (OSTARIOPHYSI, CHARACIDAE)

BY
KEITH NELSON

UNIVERSITY OF CALIFORNIA PRESS
BERKELEY AND LOS ANGELES
1964

UNIVERSITY OF CALIFORNIA PUBLICATIONS IN ZOÖLOGY
ADVISORY EDITORS: JOHN DAVIS, P. R. MARLER, R. I. SMITH
Volume 75, No. 2, pp. 59–152, 16 figures in text
Approved for publication March 20, 1964
Issued December 2, 1964
Price, $2.00

UNIVERSITY OF CALIFORNIA PRESS
BERKELEY AND LOS ANGELES
CALIFORNIA

◇

CAMBRIDGE UNIVERSITY PRESS
LONDON, ENGLAND

PRINTED IN THE UNITED STATES OF AMERICA

CONTENTS

Introduction
    General background 59
    Acknowledgments 60
    Materials and methods 60
Taxonomy and Morphology
    Descriptions of the species 63
    A comparative study of glandulocaudine morphology 69
Ecology and Distribution
    Habitat, predation, and food 75
    Breeding season 79
    Distribution 80
Non-Social Behavior
    Locomotion and orientation 81
    Feeding movements 82
    Other non-social behavior 83
    Spawning 84
Elements of Social Behavior
    Actions performed by both sexes 85
    Actions performed only or predominantly by males 91
    Actions performed only or predominantly by females 95
    Actions performed jointly by male and female 96
    Defining an action 97
Patterns of Social Behavior
    Some inferences and definitions 98
    Social interactions in non-courting animals 101
    Courtship in *Corynopoma* 110
    Courtship in *Pseudocorynopoma* 112
    Courtship in *Coelurichthys* 116
    Courtship in *Glandulocauda* 120
Discussion
    Non-social behavior 121
    Aggregation 122
    Individual recognition 124
    Systematics and phylogeny 126
    The evolution of internal fertilization 130
    Temporal patterning in glandulocaudine behavior 136
    The evolution of behavioral acts and their patterns in the
        Glandulocaudines 141
Summary and Conclusions 146

BEHAVIOR AND MORPHOLOGY IN THE GLANDULOCAUDINE FISHES
(Ostariophysi, Characidae)

BY

KEITH NELSON

INTRODUCTION

General Background

The Characidae are a large family of primitive teleosts inhabiting the freshwaters of Africa and the New World. Eigenmann (Eigenmann and Myers, 1929) has separated the genera *Corynopoma, Diapoma, Pterobrycon, Pseudocorynopoma, Gephyrocharax, Microbrycon, Hysteronotus, Landonia, Glandulocauda,* and *Mimagoniates* (including *Coelurichthys*) from the Characinae (Tetragonopterinae) as the subfamily Glandulocaudinae. Although Berg (1947), Weitzman (1960a) and others do not make this distinction, some of the reasons which Eigenmann gives for creating this subfamily make interesting reading for the student of animal behavior.

> Of greatest interest is the *ability* to develop sexual dimorphism. In the majority of the Characins the differences between the male and the female are not great. In the Glandulocaudinae, however, the males are frequently quite different from the females, and of *particular interest is the fact that this difference sometimes appears in one organ and sometimes in another*. The opercle in the males of one species is excessively modified; some of the scales in another; the fins in still others; and generally there is a glandular pouch with specially modified scales on the caudal fins of the males of all forms. In many of the species the lower caudal fulcra of the males are modified and may be separated as a spur from the rest of the fin. (Eigenmann and Myers, 1929, p. 465).

Subsequently, the aquarium literature has contained accounts of internal fertilization in several members of this group, which are in this respect quite unusual among the Characidae.

The present study was begun as a survey of the behavior of this remarkable group, largely in the hope of finding clues which might bear upon the origin and behavioral concomitants of internal fertilization in fishes.

Ethological and comparative psychological interest in teleost behavior has centered upon those groups which are highly territorial, some exhibiting striking patterns of parental care, and on live-bearing cyprinodonts. There have been only scattered studies of species which are neither live-bearing nor territorial.

From this imbalance in interest one might get the impression that territoriality and parental care represent the norm among egg-laying teleosts as among birds. They may just as likely be the exception.

More importantly, many of the generalizations in ethology are based in large measure upon observations of courtship in seasonally active, territorial fishes and birds (for examples, see Baerends, 1958); it is important to know whether they apply equally well to continuously reproductively active, non-territorial species.

Two such concepts are the distinction between appetitive and consummatory reproductive behavior and the idea of a courtship sequence as an alternating chain of actions, each of which is released by a preceding action performed by the partner (Tinbergen, 1951). The application of these and other motivational concepts to most glandulocaudine behavior was found to be difficult and dangerous and ultimately, perhaps, unnecessary.

Unfortunately, the literature on the group is scant and contradictory, and it was found necessary to conduct a morphologic and taxonomic investigation concurrently with the behavioral study. However, in addition to the support the former study gave to the behavioral conclusions, it uncovered hitherto unnoticed morphological specializations which further emphasize the remarkable nature of the Glandulocaudini.[1]

ACKNOWLEDGMENTS

This investigation was conducted during tenure of a National Science Foundation graduate fellowship, with partial support from N. S. F. grant G-13082. Field work in South America was made possible by a grant from the Associates in Tropical Biogeography. I am grateful to T. F. H. Publications and to The Aquarium Publishing Company for permission to quote from their publications.

I am especially indebted to Dr. Peter Marler for his guidance, help and patience, and to Drs. Frank A. Beach and Donald M. Wilson for much valuable advice. I am grateful to Dr. George Barlow for his valuable and painstaking critique of the manuscript. Discussions with Drs. Howard A. Bern and Jack L. Price and Mr. John Hopkirk gave rise to many of the ideas contained herein. I am especially happy to acknowledge the generosity of Dr. Price, whereby I was enabled to spend a month on his estate on the Maracas River in Trinidad, to see records of many collections made by him, and to use the facilities of the Bamboo Grove Experimental Fish Farm, whose personnel were very helpful.

Drs. George Dahl and Federico Medem showed me many courtesies in Colombia, and I am also indebted to Dr. Harald Schulz for supplying the Southern Brazilian material of this study, and much information as well. Dr. Stanley Weitzman has checked some of the taxonomic material, and has listened with patience to some of my theories.

MATERIALS AND METHODS

Several examples of *Corynopoma riisei* Gill were obtained from a dealer in the fall of 1958, and the stock was replenished from different sources twice in 1960. These fish, perhaps as a result of aquarium inbreeding, suffered heavily from diseases, and until September, 1961, the quantitative study of this species was largely limited by the amount of material available. In that month healthy stocks of *Corynopoma riisei* were collected from several localities in the Caroni drainage of Trinidad. These stocks were kept distinct and were used exclusively after September, 1961.

In June of 1962, through the courtesy of Harald Schultz of São Paulo and his brother Dr. A. Viggo Schultz of Porto Alegre, stocks of five other Glandulocaudines were obtained from Karl-Heinz Stegemann of Tropical Aquario Ltda., São

---

[1] Noun and adjectival endings for subfamilies (-inae, -in) and tribes (-ini, -ine) follow the usage of Weitzman (1960b).

Paulo. These, with their localities, were as follows: *Glandulocauda inequalis* Eigenmann and *Pseudocorynopoma doriae* Perugia, tributary of Guahyba Bay, Rio Grande Do Sul; *Ps. heterandria* Eigenmann and *Coelurichthys microlepis* (Steindachner), Iguape, São Paulo; and *C. tenuis* Nichols, vicinity of Santos, São Paulo. Unfortunately, the stocks of *G. inequalis* and *Ps. heterandria* each contained only one mature male; that of *Ps. doriae* included only one female. Studies on these species were restricted by these limitations in numbers.

For comparative purposes, stocks of other characids were maintained at various times. These included *Astyanax bimaculatus, Moenkhausia oligolepis, Knodus* sp., *Hemibrycon dentatus, H.* sp., *Hemigrammus unilineatus, Bryconamericus* sp., *Triportheus* sp., *Prionobrama* sp., *Aphyocharax rubripinnis*, and others.

Fish were kept, singly and in groups, in 5-, 7-, 15-, and 30-gallon aquaria, all heated to 24–25° C., with the exception of certain tanks containing *G. inequalis* and *Ps. doriae* which were maintained at 18–22° C. All tanks were planted with one or more of the following: *Elodea, Sagittaria, Heteranthera, Fontinalis,* and *Nitella*. In addition, *Lemna* covered much of the surface of some tanks. Only the walls of tanks through which observations were made were kept free of the inevitable algae. The natural lighting was supplemented by 25 to 75 watt overhead bulbs. No attempt was made to artificially control pH or hardness, but it was found that tanks would maintain a fairly stable pH for months at a time. The six 15-gallon observation aquaria had pH's ranging from 6.0 to 7.0; *C. tenuis* in particular was always observed in aquaria with an acid balance.

After September, 1960, all tanks contained small catfish of several callichthyid genera; these were removed when they interfered with observations. The fish were maintained on live brineshrimp (*Artemia*) and dried food; in certain experiments live *Drosophila* were introduced into the space between the aquarium cover and the surface of the water.

Observations on the development of social and feeding behavior in the young up to the time of sexual differentiation were made of *C. riisei, C. microlepis,* and *G. inequalis* in groups of ten to two hundred young in 15-, 30-, and 50-gallon tanks. In addition, the interactions between juveniles of different age classes and between these and adults were observed in *C. riisei*.

The nature of different encounter situations which were set up will be described in later sections.

Timed records of the behavior of *Corynopoma riisei* were made using a tape recorder; later, records of encounters in all species but *Ps. heterandria* were made using an Esterline-Angus 20-channel operation recorder coupled to a keyboard designed and built by Mr. George Hersh. A further description of these recording procedures will be found under the heading "Defining an Action."

Photographs in black and white were taken of various activities and in some cases were keyed to timed records made simultaneously of encounters taking place in a special 15-gallon observation tank. Black-and-white motion pictures were made of aggressive and courtship encounters in the two species of *Coelurichthys*.

Attempts were made to persuade male *Corynopoma* to respond to an accurate wax model of a female; these were without success. The results of several interspecific encounters will be reported in later sections.

In August, 1961, several days were spent in the Colombian Llanos between Villa Vicencio and Puerto López, searching without success for *Corynopoma*. Following this a month was spent in Trinidad. While the turbid waters of the rainy season precluded direct observation of *Corynopoma* behavior in the field, by collecting at many locations a fair picture was built up of the ecology of that and other species on that island. In addition, several aquaria and concrete tanks were available on the experimental fish farm at Bamboo Grove, and it was possible to study the relations of *Corynopoma* to other species under at least seminatural conditions.

I examined a large series of records of collections made by J. L. Price in May–August, 1954 (Price, 1955), and December, 1957–March, 1958. Including the collections made by Dr. Price and me in 1961, a total of 284 collections from all parts of the island were recorded for analysis.

In the morphological studies reported in the next section, the following material was examined. Lengths given are standard lengths; asterisks denote material which was cleared and stained according to the method of Clothier (1950).

The following abbreviations have been used:

BM	British Museum (Natural History)
CAS	California Academy of Sciences
CAS/IUM	Indiana University Museum Collection purchased by the California Academy of Sciences
SU	Stanford Museum of Natural History
USNM	U. S. National Museum

Acrobrycon ipanquianus (Cope), 1 male*, 102 mm., several others, CAS/IUM 16055, Torontoi, Peru.

Argopleura chocoensis (Eig.), 1 male*, 45 mm., "Duplicate of Paratype," CAS/IUM 12939, Istmina, Colombia.

Argopleura diquensis (Eig.), 1 male, 48 mm., paratype, CAS/IUM 12820, Soplaviento, Colombia.

Argopleura magdalenensis (Eig.), 1 male, 53 mm., CAS/IUM 12824, Rio Cauca at Cali, Colombia.

Bryconamericus sp., 1 male*, 36 mm., aquarium specimen.

Coelurichthys microlepis (Steind.), many*, to 55 mm., Iguape, São Paulo.

Coelurichthys tenuis Nichols, 2 males*, to 29 mm., several females*, Santos, São Paulo.

Compsura gorgonae (Evermann and Goldsborough), 2 males*, 23 and 28 mm., SU 48851.

Corynopoma riisei Gill, many, to 50 mm., 1 male*, 37 mm., Bridge 1/2 Carapo Rd., Trinidad.

Gephyrocharax atricaudata (Meek and Hildebrand), several, CAS 6820, Rio Cocoli, C.Z.; 1 male, 34 mm., SU 50909, Gamboa, Panama.

Gephyrocharax chocoensis Eig., 1 male*, 46 mm., several others, CAS/IUM 17292, Ixiamas, Bolivia.

Gephyrocharax melanocheir Eig., 1 male*, 30 mm., several others, SU 50376; several, SU 49495; one male, SU 50387, Rio Magdalena system near La Dorada, Colombia.

Glandulocauda inequalis Eig., 1 male*, 31 mm., aquarium specimen, vic. Rio Jacui, Rio Grande Do Sul; 1 male, 42.6 mm., USNM 94310, Porto Alegre, Brazil.

Hemibrycon sp., 1 male*, 50 mm., aquarium specimen.

Knodus sp., 1 male*, 46 mm., aquarium specimen.

Landonia latidens Eigenmann and Henn, 1 male, 37 mm., several others, paratypes, CAS/IUM 13100, Vinces, Ecuador.

Mimagoniates barberi Regan, 2 syntypes, including lectotype designated by L. Schultz (1959), BM 1907.10.22.5, Arroyo Yaca, Estación Caballero, Paraguay.

Paragoniates alburnus Steind., 1 female*, 55 mm., 1 male*, 60 mm., SU 50670, Tres Esquinas, Colombia.
Phenacobrycon henni (Eig.), 1 male, 36 mm., several females, paratypes, CAS/IUM 13102, Vinces, Ecuador.
Planaltina myersi Böhlke, 1 male, type, SU 18636, 37 mm., Planaltina, Goyaz, Brazil.
Triportheus sp., several, juveniles, to 24 mm., aquarium specimens.
Tyttocharax rhinodus Böhlke, 2, paratypes, SU, vic. Tingo Maria, Huanaco, Peru.
Xenurobrycon macropus Myers and Miranda-Ribeiro, 2, SU 40764, paratypes.

In addition, several collections of material belonging to the *Mimagoniates-Coelurichthys* complex were compared with the material of the present study. These bore the U.S. National Museum numbers 94117, 177701, 177703, and 177820.

Drawings of caudal glands of selected specimens were made using a binocular microscope. In addition, the caudal peduncles of mature males of *Coelurichthys tenuis* and *Corynopoma riisei*, and the anal fin of the former, were fixed in Bouin's fluid with acetic acid, sectioned at 10μ and stained with a modified Masson's fluid, through the courtesy of Dr. Howard Bern.

TAXONOMY AND MORPHOLOGY

DESCRIPTION OF THE SPECIES

All species studied possessed in both sexes the following combination of features, which may be referred to as the glandulocaudine habitus:

Premaxillary teeth notched, in two series, mandibulary-maxillary teeth in a single series.
Origin of dorsal... distinctly behind the middle of the body... anal short or of moderate length; mouth very oblique, the lower jaw quite or nearly entering the profile; pectorals large, falcate, reaching beyond origin of ventrals, frequently to the anal; profile from dorsal to snout nearly straight, the ventral profile from chin to ventrals arched; second suborbital usually covering the entire cheek. (Eigenmann and Myers, 1929, p. 466)

In addition, each species had its own peculiarities, which are described in the following paragraphs. However, first it must be said that there is currently much confusion regarding the specific and even generic identity of several of the fishes of the present study, and accordingly, a brief history of the nomenclature of the three genera *Mimagoniates*, *Coelurichthys*, and *Glandulocauda* must be given in order that the present usage may be understood. The following remarks are in no way intended to be a taxonomic revision of the genera involved, and accordingly synonymies and quantitative descriptions will be omitted from the species descriptions given below.

In 1876, Steindachner described *Paragoniates microlepis* from creeks near Rio de Janeiro. Regan (1907) erected a new genus for a similar fish from Paraguay, *Mimagoniates barberi*. Evidently unaware of Steindachner's and Regan's species, Miranda-Ribeiro (1908) erected another genus to receive a similar species, *Coelurichthys iporangae*, from southeastern Brazil. In 1911, Eigenmann erected the genus *Glandulocauda*, and placed within it the three southeastern Brazilian species, *G. melanogenys*, *G. inequalis*, and *G. melanopleura*. Nichols (1913) described two more fishes which he considered specifically distinct from one another and from *Coelurichthys iporangae* Miranda-Ribeiro, and which he named *Coelurichthys lateralis* and *C. tenuis*. No locality data were available on these two species.

In his monograph on the American Characidae (Eigenmann and Myers, 1929), Eigenmann synonymized *P. microlepis* Steindachner and *C. iporangae* Miranda-Ribeiro under the name *Coelurichthys microlepis*, pointing out that the type species of *Paragoniates* belonged to a different group. Myers, in revising Part 5 of that monograph for publication, inserted a note to the effect that *Coelurichthys* was a synonym of *Mimagoniates* Regan, which had priority (ibid., page 464, page 493). At the same time Myers stated that the type of *C. lateralis* was a female, and that of *C. tenuis* an "emaciated" male, of *M.* (= *C.*) *microlepis* (ibid., page 492). He compared Eigenmann's material of *M.* (= *C.*) *microlepis* with specimens (USNM 94117) which Norman had pronounced to be specifically identical with Regan's types of *M. barberi*, and retained with hesitation the distinction between *M. microlepis* and *M. barberi* (ibid., page 493).

Böhlke (1958a, p. 43) retained the two genera *Mimagoniates* and *Glandulocauda* as they had previously been described (Eigenmann and Myers, 1929), and distinguished them in a key in part as follows:

e ... on males of *G. melanogenys*, the anal hooks are relatively smaller and there are a number of them on each of the anteriormost 7 to 9 branched anal rays, the more posterior rays each bearing only a single hook; also, while there are scales with modified shapes on the base of the caudal, these lie flat and are not bent, furrowed or twisted to form a complex with any depth. (*Glandulocauda* Eigenmann)

ee ... males with the anal hooks relatively larger and only one per ray on all the anal rays bearing hooks; a complex thickened structure at mid-base of caudal, formed of peculiarly bent and twisted scales. (*Mimagoniates* Regan)

Recently a paper (L. Schultz, 1959) appeared which has thrown the nomenclature of these genera into further confusion. Among other things, in this revision of the two genera *Mimagoniates* and *Glandulocauda* Schultz has combined them under the name of *Mimagoniates*, has considered *G. inequalis* Eigenmann and the *M. microlepis* (Steind.) in part of Eigenmann to be specifically identical, and has placed in *M.* (= *G.*) *inequalis* the material (USNM 94117) which Norman stated to be *M. barberi* Regan.

An examination of two of Regan's syntypes and of other material now at the U.S. National Museum, and their comparison with the Brazilian material, live and preserved, of the present study, has led me to the following tentative conclusions. The Brazilian material of the present study represents three species which on behavioral and morphological grounds are quite distinct from one another. None of them can be considered to be specifically identical with the two syntypes of *M. barberi* Regan which were examined. One is evidently the same as *Glandulocauda inequalis* Eigenmann, and will be so considered here. Another is identical with *C.* (= *M.*) *microlepis* (= *C. iporangae*), and the third is identical with at least some of the specimens in USNM 94117, and appears to most closely resemble Nichols' *C. tenuis*.[2] The latter two groups are closely related, and quite distinctive. If they are retained in *Mimagoniates*, as *M. microlepis* and *M. tenuis*, there appears to be no alternative but to include *Glandulocauda* within *Mimagoniates* (as did L. Schultz, 1959), for *inequalis* is morphologically closer to *M. barberi* Regan than are *microlepis* and what is here considered to be *tenuis*. The alternative to lumping

---

[2] Subsequent examination of Nichols' type (AMNH) of *C. tenuis* confirmed this identification. The relationships of *C. lateralis* are still problematical.

all these under *Mimagoniates* is to revive *Coelurichthys* Miranda-Ribeiro to receive *microlepis, tenuis* and *lateralis,* and at the moment this appears to be the safest procedure. Whether *C. lateralis* Nichols contains females specifically identical to *C. microlepis* or to *C. tenuis* is difficult to decide; as males in this group are more easily identified than are females, *C. tenuis* Nichols will be used here. However, later investigation may show that *C. lateralis* and *C. tenuis* indeed refer to the same species, and if so, *C. tenuis* must be replaced by the prior name *C. lateralis*.

The names for species in these genera used in the present study, then, will be *Glandulocauda inequalis* Eigenmann 1911, *Coelurichthys microlepis* (Steind.) as understood by Eigenmann (Eigenmann and Myers, 1929), and *Coelurichthys tenuis* Nichols 1913. Otherwise the nomenclature used will follow Eigenmann and Myers (1929) and Böhlke (1958a).

Corynopoma riisei Gill (fig. 1, *a*).—In both sexes the caudal peduncle is prolonged over the usual glandulocaudine condition, and in the male it is considerably less deep at the level of the last anal ray, and widens posteriorly. Both sexes possess relatively small pelvic fins.

Sexual dimorphism is extreme. The differences in median fins shown in the illustration need little comment. Not well shown are a prolongation of the last few anal rays, and the presence of a spur formed of the lower caudal fulcra (fig. 4, *f*). Not shown are a patch of retrorse hooks on anal rays 3 through 12 or 13, and similar hooks on the pelvic fins and on the lower lobe of the caudal (fig. 4, *f*). Perhaps the most remarkable feature is the opercular extension or paddle, enlarged at the tip into a plaque. This latter is variously shaped but generally similar to the one shown. The anal in the male possesses a sheath of scales at its base.

There is some sexual dimorphism in color pattern as well. Both sexes are a nondescript olivaceous yellow above and creamy white below, and are iridescent in reflected light. The top and bottom rays of the caudal fin in the female are translucent whitish yellow, the male's caudal extension is likewise whitish yellow, and the remainder of the fins in both sexes is almost colorless and transparent. Both sexes possess a dusky humeral spot, behind which a gray stripe runs to the base of the caudal, widening posteriorly. This latter is faint in the male, and instead the caudal peduncle bears a large, indistinctly defined gray blotch, in the center of which is a punctate black spot. There is a similar spot just forward of the pelvic fins, another on the paddletip, and a crescentic mark over the base of the last few anal rays. All of these become darker during courtship, and the ventral dot will be referred to in later sections as the "belly spot."

The caudal gland of this and the other species will be described at the conclusion of this section. At this point it may be mentioned that *Pterobrycon landoni* Eigenmann (illustrated in pl. 67, fig. 4 of Eigenmann and Myers, 1929) possesses an organ remarkably similar to the opercular extension of *Corynopoma*, except that it is a modified scale—specifically the fifth up from the lateral line in the seventh row.

Pseudocorynopoma doriae Perugia (fig. 1, *b*).—In both sexes the most prominent feature is the keeled ventral surface. The pectoral fins are set some distance up from the lower edge of the body and in the male overlap the anal fin when pressed back. This and the following species are more compressed laterally than are the

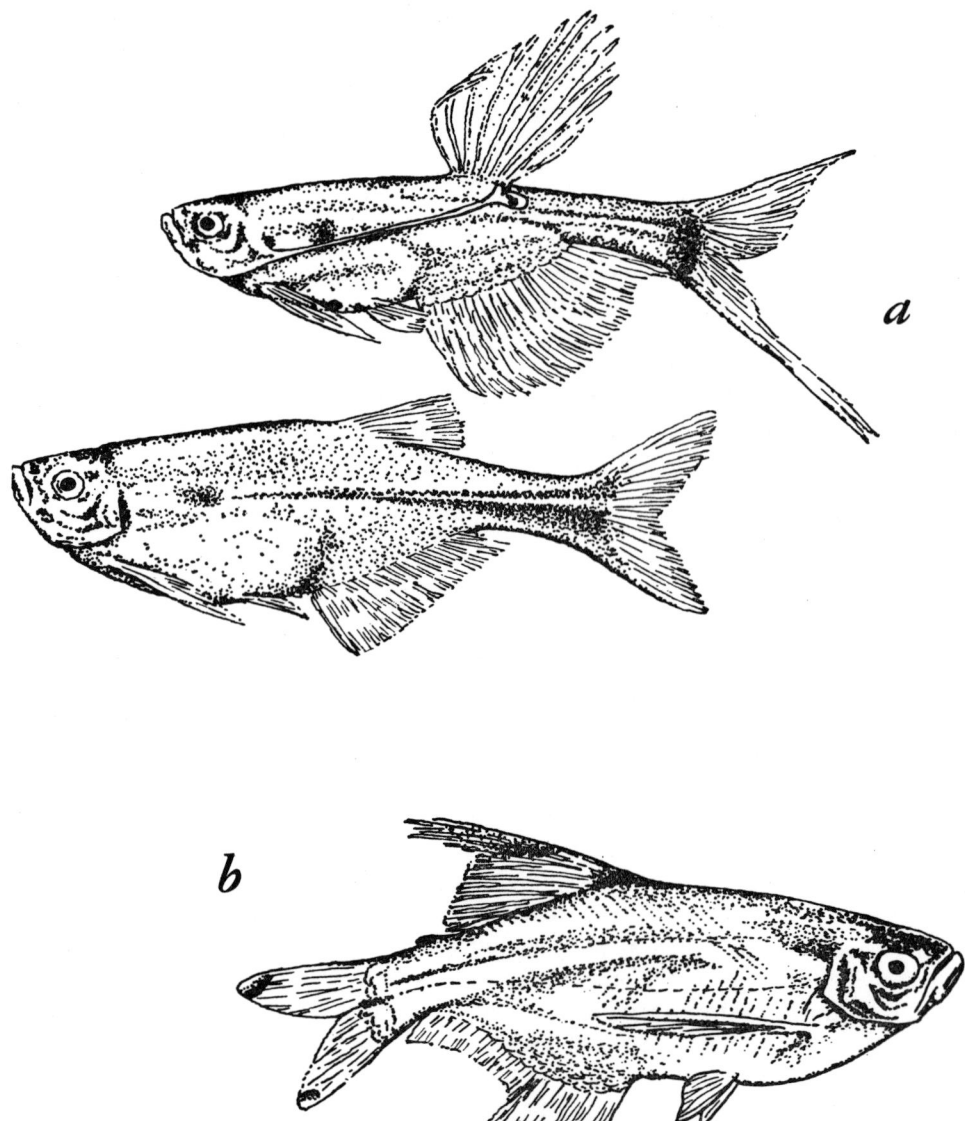

Fig. 1. *a. Corynopoma riisei;* male, above, female below, × 1.8.
b. Pseudocorynopoma doriae; male, × 1.6.

other species studied; in most specimens the first dorsal ray is inserted directly above the first anal ray, and the pelvic fins are small. In both sexes the body color is a warm silvery white, and there is an ill-defined black spot extending onto the caudal fin. The fins are hyaline except for the tips of the dorsal and caudal. Both lobes of the caudal are tipped with a tricolor black, red, and white dot. The tip of

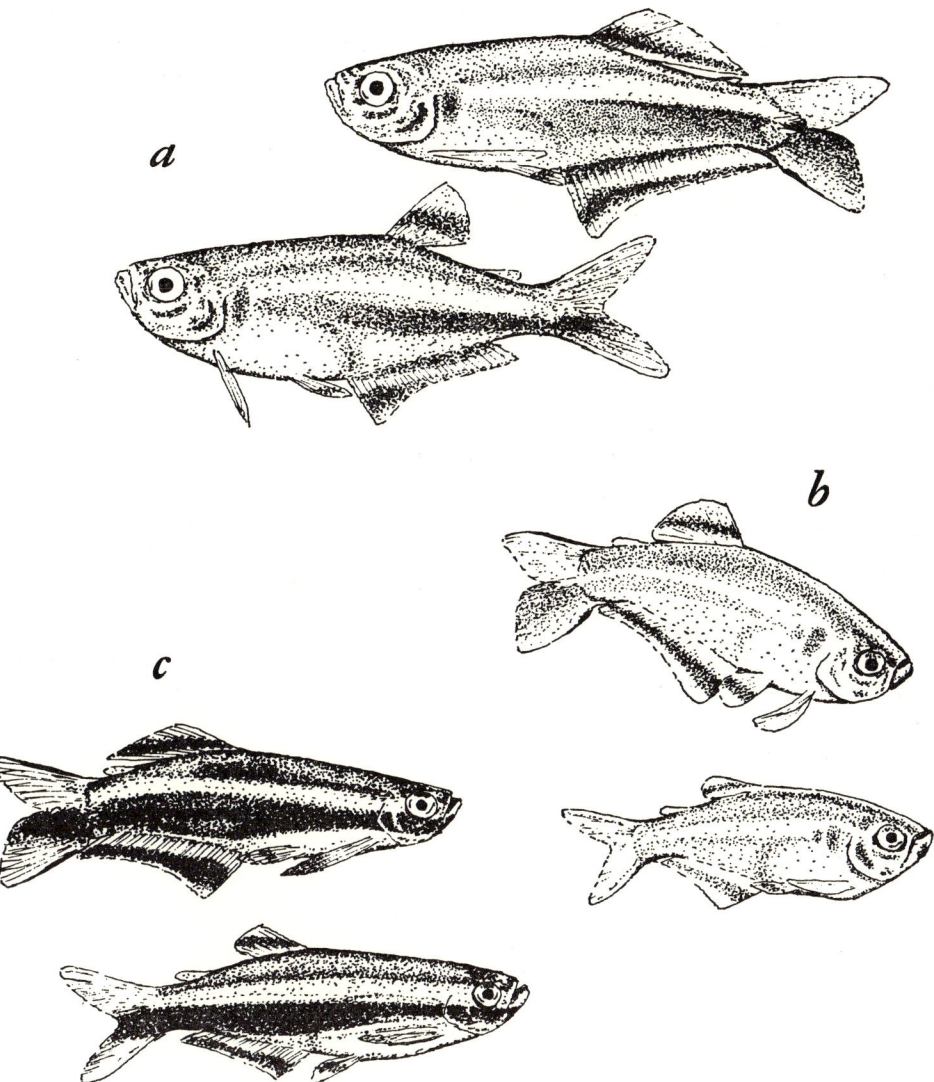

Fig. 2. *a. Coelurichthys microlepis*; male above, female below, ×1.4. *b. Glandulocauda inequalis*; male hovering above female, from colored slide, ×1.6. *c. Coelurichthys tenuis*; male above, ×1.6.

the dorsal in the female is crossed by an oblique dusky bar, and this shades into translucent white at the upper edge. In the male, the same area is dusky pink, and the adipose fin and base of the dorsal are gray to black.

Sexual dimorphism is confined to the length and shape of the fins, to the caudal gland, and to a prominent anal sheath in the male. The dorsal in the male is over three times as long as in the female, and the first four rays are one and one-half times the length of the others. The anal fin is divided into three lobes, the rays of the first lobe three times, and those of the middle twice the length of the rays in the posterior lobe. The rays of the middle lobe bear small retrorse hooks; in the material at hand there were no hooks on the pelvic or caudal fins. Dimorphism also exists in the length of the pectoral fins; these did not have the black tips spoken of in Eigenmann and Myers (1929). A good color illustration may be found in A. V. Schultz (1962).

Pseudocorynopoma heterandria Eigenmann (not illustrated).—The foregoing points apply to this species as well, with the following modifications. The keeled ventral profile is more angular than in *Ps. doriae*, that is, the profile forward of the pectorals is steeper.

In both sexes the caudal spot is a longitudinal, sharply defined oval, larger than in *Ps. doriae*; this feature is not shown in Eigenmann's illustration (Eigenmann and Myers, 1929, pl. 83, fig. 1). In the male the anal sheath covers only the base of the forward part of the fin, which has only two lobes. The dorsal fin is only slightly dimorphic. All fins in both sexes are hyaline, except for a small black-and-white dot at the tip of the upper caudal lobe only.

Glandulocauda inequalis Eigenmann (fig. 2, b).—This is a short, chunky species, olivaceous above and silvery white below. There is a vertically elongate humeral spot and some indication of a thin lateral stripe, widening posteriorly. The upper half of the eye is red; this is more intense in the male.

There is considerable sexual dimorphism. Males are larger, and their bodies are somewhat arched at the level of the pelvic fins, so that the dorsal profile describes a pronounced curve. This arching is present but more gentle in the female. The anal fin and lower caudal lobe are fuller in the male. In the males there are retrorse hooks on the pelvics, but none on the caudal. On the anal there are two hooks on each of the first three rays and one each on the next four. The caudal peduncle is deeper in the male. Indistinct dusky yellow bars pass obliquely across the pelvic, anal and dorsal fins, leaving whitish areas distally to them. The caudal is tinged with yellow, extending forward along the upper fulcra. The lower lobe is edged with black, and has at its top margin a broad dusky band. The lower lobe is notched on its ventral edge where it joins the caudal peduncle. The markings become darker during aggressive behavior and against a dark background. In the female the markings are less distinct. The male is excellently shown in the upper fish of the color photograph on page 9 of L. Schultz (1959), labelled *"Mimagoniates microlepis."*

Coelurichthys microlepis (Steindachner) (fig. 2, a).—The shape of this larger species (see "Materials and Methods" for sizes) is similar to *G. inequalis*, but somewhat more elongate and fusiform. The color is light olivaceous brown above to yellowish silver below, with an indistinctly demarked, broad, bluish lateral stripe extending from the gill cover onto the lower caudal lobe, becoming darker and somewhat broader posteriorly. Centered on this stripe anteriorly is an indistinct humeral blotch. As in *G. inequalis*, oblique dusky stripes cross the pelvics and dorsal; in contrast to *G. inequalis*, the dark area on the anal follows the ventral outline of the fin posteriorly. The dark area is crossed by an oblique iridescent light blue stripe, seen only in certain lights; the base of the fin is goldish, becoming dark in aggressive encounters. The outline of the anal is emarginate; in *Glandulocauda* it is straight. Both caudal lobes are yellowish.

Sexual dimorphism affects most of the same characters as in *G. inequalis*. The anal and dorsal are broader and longer in the male; the lower caudal lobe is fuller and separate from the upper. There is the same notch between the lower caudal fulcra (fig. 3, *d*). The caudal peduncle is broader. There are numerous hooks on the pelvics and on the four modified caudal rays (see next section), where they point anteromedially. There is one large retrorse hook on each of the first nine branched anal rays. A deep-lying, iridescent greenish-yellow stripe above the lateral stripe is sometimes more evident in males.

In both sexes the dark fin markings become darker during aggressive encounters, and in courtship the background color of the male turns ruddy. The female's fin markings are, under equivalent conditions, less distinct than in the male; on the other hand, her bluish lateral stripe is generally more distinct. The lower fish in the illustration on page 9 of L. Schultz (1959) is probably a young male of *C. microlepis* (Steindachner).

Coelurichthys tenuis Nichols (fig. 2, *c*).—In L. Schultz's (1959) paper, the forms here considered to be *C. tenuis* Nichols are shown in an excellent color photograph on page 8, labelled "A group of male *Mimagoniates inequalis*," and evidently considered in the text to be *M.* (=*C.*) *microlepis*.

This is a much more elongate form than either *G. inequalis* or *C. microlepis*, and has much straighter dorsal and ventral profiles. Otherwise it is similar to *C. microlepis*, and the points of morphological dimorphism hold for this species as well, with the following exceptions. In the male only the last unbranched and first three branched anal rays bear a hook, and on the middle caudal rays there are protuberances but no hooks.

The color pattern in both sexes is quite distinctive. The back is brown, somewhat darker in a stripe from above the hind edge of the gill cover and extending to below the last rays of the dorsal fin. Below this, from the snout through the eye and extending through the top half of the caudal peduncle, runs a gleaming gold (in some lights greenish) band. Below this, from the chin to the trailing edge of the lower caudal lobe, is a dark band which widens posteriorly. Below this the body is cream to light gold. The dorsal, anal, and pelvic fins are banded as in *C. microlepis*, except that the bands are deep reddish-brown to black. Proximal to these bands the dorsal and anal fins are under certain conditions deep red. The remaining parts of the fins are yellowish and in the male the caudal is bordered above and below with black.

In both sexes the lateral band and the fin markings become darker during aggressive encounters. The lateral band in particular becomes broader and an intense blue-black. During courtship the lateral band in the male becomes paler and split into two above the anal fin by a long oval of iridescent greenish yellow.

The fish described by Sterba (1959) under the name *Mimagoniates barberi* may be the same species.

A COMPARATIVE STUDY OF GLANDULOCAUDINE MORPHOLOGY

Böhlke (1954) ventured the opinion that the glandulocaudines did not form a natural group; later (1958*a*) he categorically denied that they were monophyletic in origin. Weitzman (1960a) is of the opinion that there is at the present time insufficient evidence to decide the question. Both authors indicate that a detailed study is needed of the caudal glands in the various genera.

As this question is of fundamental importance for the interpretation of the evolution of behavior in this group, it was felt that it could not be ignored in the present study. Therefore preserved material of males of ten of the thirteen genera listed by Böhlke (1958a) in his key to the glandulocaudines, two of the genera placed by Myers and Böhlke (1956; Böhlke, 1958b) in the Xenurobryconini, but considered by them to be offshoots of a glandulocaudine stock, and five other characid genera, making seventeen in all, were examined. In addition, cleared and stained material of thirteen of these seventeen genera was studied. Closest attention was paid to the morphology of the caudal skeleton and the caudal glands in the Xenurobryconini and Glandulocaudini. The results of these studies are presented in the following paragraphs and in figures 3, 4, and 5, a–b.

It is not practical to determine the histological nature of the skin of museum material, and as Dr. Howard Bern (pers. comm.) aptly pointed out, "all fish skin is glandular." Nevertheless, in the glandulocaudine and xenurobryconine genera studied, thickened areas of naked skin were present on the caudal fin which were not found in the other genera studied. By analogy with the situation in *Pseudocorynopoma doriae* (pers. obs.), *Corynopoma riisei* (Kutaygil, 1958; Weitzman, pers. comm., and pers. obs.), and *Coelurichthys tenuis* (Weitzman, pers. comm., and pers. obs.), where these areas are known to be concentrations of secretory-appearing cells, with sloughing and other signs of exocrine activity, such areas in the present paper are considered to be "glandular."

The present findings indicate that the unifying features of caudal morphology which set the Glandulocaudini and Xenurobryconini apart from other characids are the presence of these thickened glandular areas in combination with some sort of modified caudal scales. A thickened glandular area was always found at some point along the middle caudal rays (figs. 3, b, c, d; 4, a, b, c, d, e, and f; fig. 5, a, and [not illustrated] in *Planaltina* and the Xenurobryconini). Another glandular area was seen in the region of the lower caudal fulcra in *Coelurichthys* (fig. 3, c and d), *Glandulocauda* (fig. 5, a), *Acrobrycon* (fig. 4, d), *Argopleura* (fig. 4, c), *Gephyrocharax* (fig. 4, e), and *Corynopoma* (fig. 4, f). Terminal scales on the lower half of the caudal peduncle were raised to form a pouch in all glandulocaudine genera but *Pseudocorynopoma, Landonia,* and *Glandulocauda* (fig. 5, a). This was much reduced in *Mimagoniates, Coelurichthys* (fig. 3, c, and d) and *Phenacobrycon* (fig. 3, b). A pouch on the lower caudal peduncle is also present in at least two of the three genera not examined, *Pterobrycon,* which is obviously not far removed from *Gephyrocharax* and *Corynopoma* (Eigenmann and Myers, 1929), and *Hysteronotus,* which Böhlke (1958a) placed close to *Pseudocorynopoma*. Modified caudal scales extend far out onto the central caudal rays in *Mimagoniates, Coelurichthys, Glandulocauda, Landonia, Argopleura, Acrobrycon, Planaltina,* and the Xenurobryconines of the genera examined, and of those not examined, in *Hysteronotus* (Böhlke, 1958a) and *Diapoma* (Eigenmann and Myers, 1929). This extension is formed of a greatly enlarged and peculiarly sculptured scale in *Argopleura, Tyttocharax,* and *Xenurobrycon,* and in *Microcoelurus* (Miranda-Ribeiro, 1939;

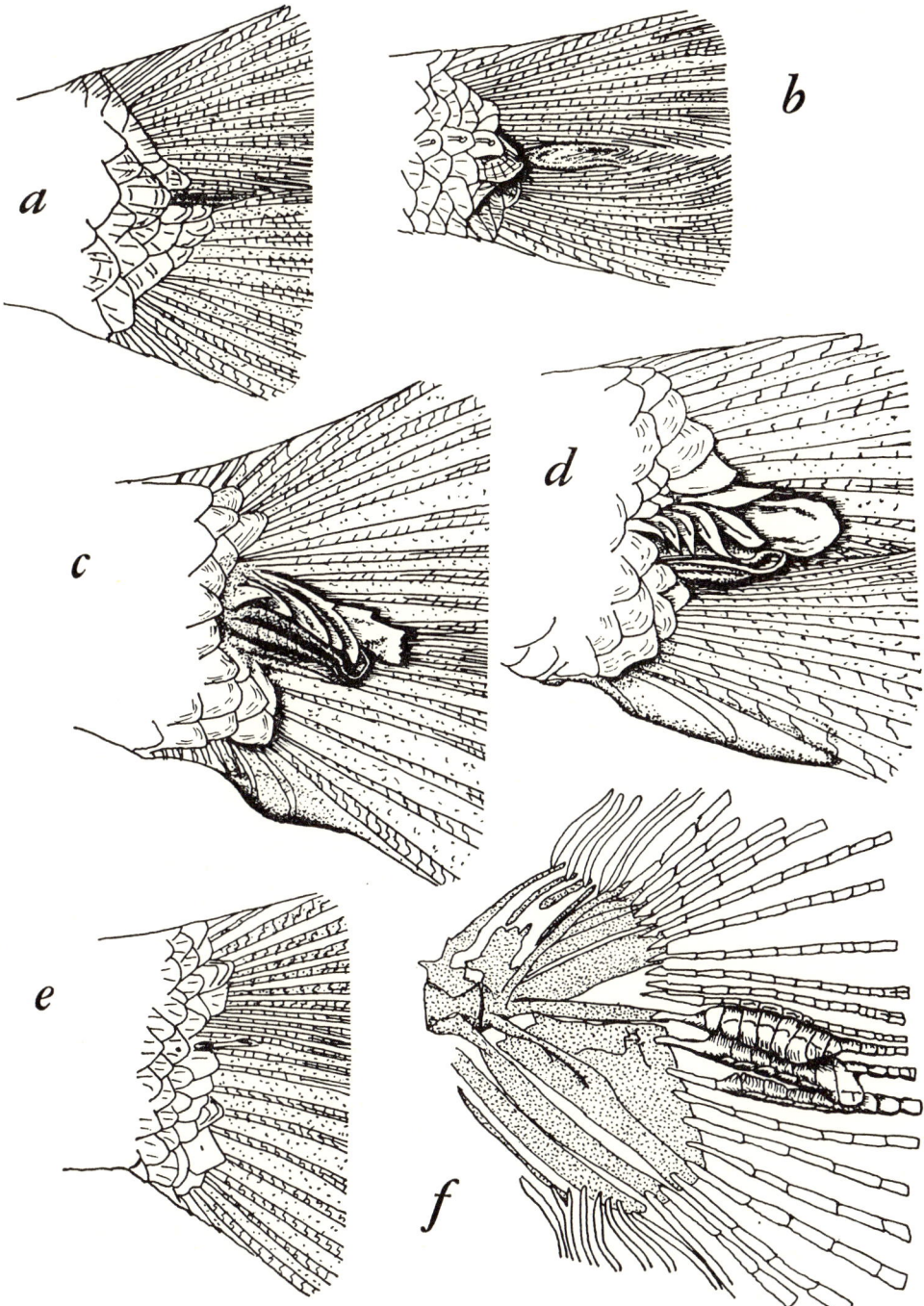

Fig. 3. Caudal Morphology: *a. Compsura gorgonae;* male, 23 mm., SU 48851. *b. Phenacobrycon henni;* male, 36 mm., CAS/IUM 13102. *c. Coelurichthys tenuis;* male, 27 mm., not catalogued. *d. Coelurichthys microlepis;* male, 53 mm., not catalogued. *e. Paragoniates alburnus;* male, 60 mm., SU 50670. *f. Coelurichthys tenuis;* male, 29 mm., cleared, not catalogued.

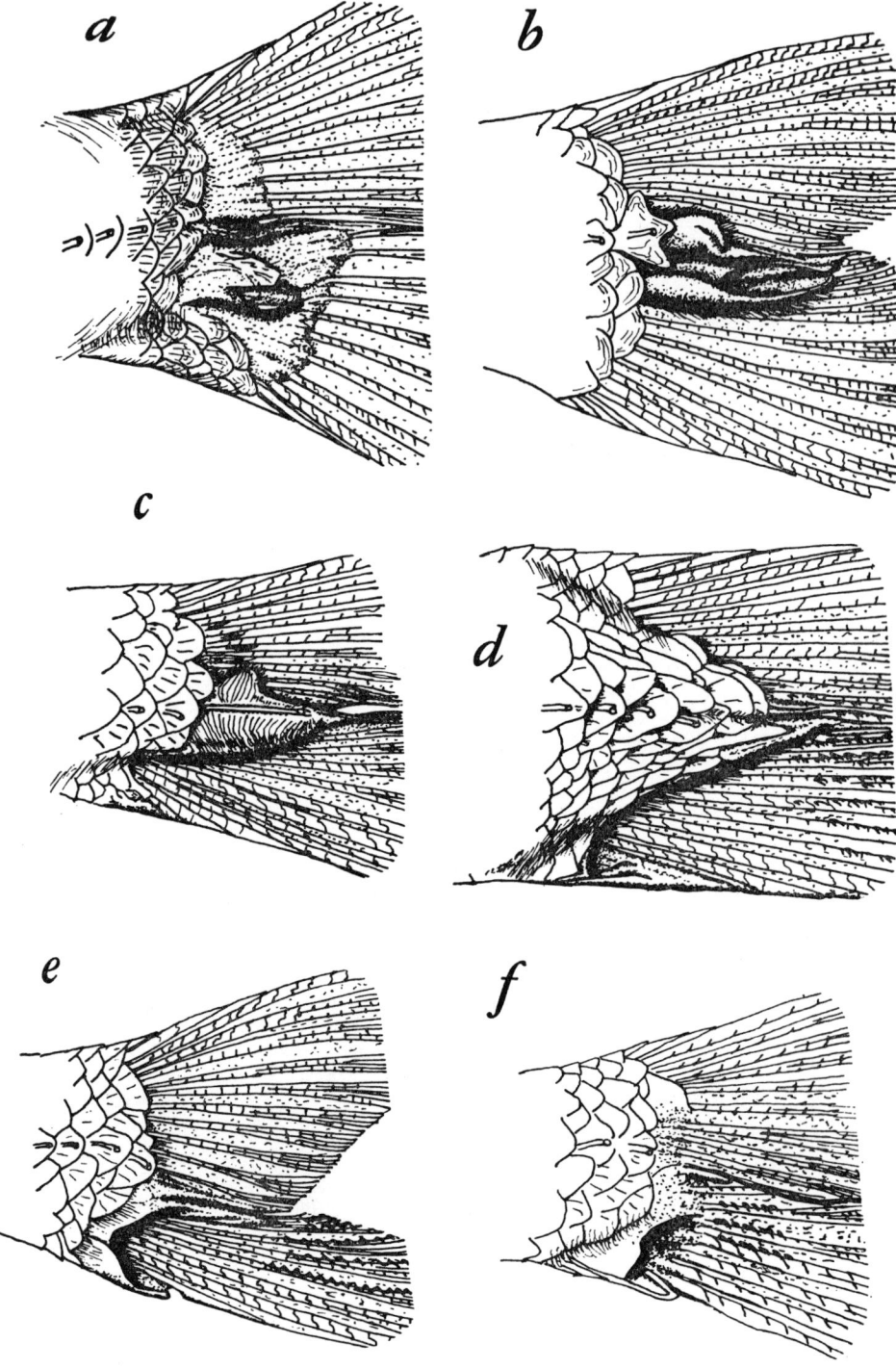

Fig. 4. Caudal Morphology: *a. Pseudocorynopoma doriae;* male, 50 mm., not catalogued. *b. Landonia latidens;* male, 37 mm., CAS/IUM 13100. *c. Argopleura chocoensis;* male, 45 mm., CAS/IUM 12939. *d. Acrobrycon ipanquianus;* male, 102 mm., CAS/IUM 16055. *e. Gephyrocharax chocoensis;* male, 46 mm., CAS/IUM 17292. *f. Corynopoma riisei;* male, 37 mm., not catalogued.

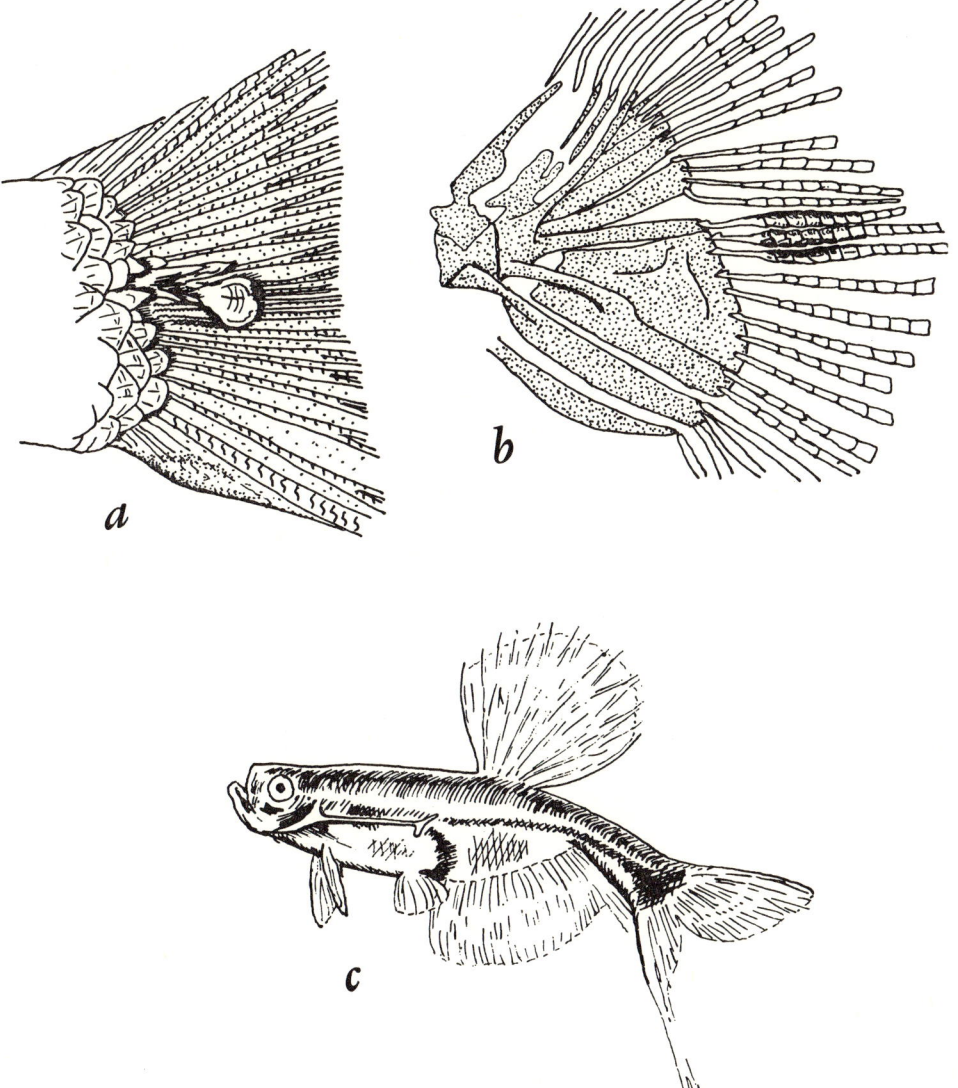

Fig. 5. Caudal Morphology: *a. Glandulocauda inequalis;* male, 31 mm., not catalogued. *b.* Same specimen, cleared (both Caudal Morphology). *c.* Male *Corynopoma riisei,* Yawning. Note extended paddle. (From photograph.)

not examined). A scale near the middle caudal rays rolled into a tubular structure is present in *Pseudocorynopoma, Landonia,* and *Hysteronotus* (Böhlke, 1958a; not examined). *Landonia* otherwise resembles *Phenacobrycon* in general appearance (Böhlke, 1954, and pers. obs.); both genera are from the same locality in Ecuador (Eigenmann and Myers, 1929).

A similar tubular structure was seen in both species of *Coelurichthys* (fig. 3, *c* and *d*) but as Mr. John Hopkirk pointed out to me, here it is formed of the peculiarly modified central caudal rays 10, 11, 12 and 13, counting from the top (fig. 3, f); this structure is bilaterally symmetrical in all specimens examined. The caudal skeleton of *Coelurichthys* possesses several other peculiarities. The split in the caudal fin is between rays 9 and 10, rather than between rays 10 and 11, a situation which has been observed in no other characid. In *Gephyrocharax chocoensis* the split occurs between rays 12 and 13; no such shift occurs in *G. melanocheir*. The only other recorded case known to this author in which the generalized characid formula of 10:9 has been modified is that of some species of *Poecilobrycon*, belonging to a radically different group (Weitzman, 1960b). Associated with this shift in *Coelurichthys* was a large, triangular gap between hypurals 4 and 5, leaving rays 8, 9 and 10 free of hypural support (fig. 3, f). A less pronounced gap between hypurals 4 and 5 was seen in cleared material of *Acrobrycon*, *Corynopoma*, and *Glandulocauda* (fig. 5, b), and among the non-glandulocaudine genera *Bryconamericus*, *Knodus*, one specimen (the larger) of *Compsura gorgonae*, *Hemibrycon*, and *Paragoniates;* Weitzman (pers. comm.) states that it is not uncommon among the Characidae.

Böhlke (1958a) in his key quoted above, separated his *Mimagoniates* and *Glandulocauda* on the basis of the single row of anal hooks and complex, thickened caudal gland in the former genus. Böhlke's "peculiarly bent and twisted scales" in the caudal gland of his *Mimagoniates* are the modified fin rays of *Coelurichthys microlepis* and *C. tenuis* referred to above. Only a slight thickening of the middle caudal rays is present in Regan's syntype of *Mimagoniates barberi*, and the rays are similarly only slightly thickened in the mature male of *G. inequalis* (fig. 5, b). Again, both *M. barberi* Regan and *G. inequalis* have more than one hook on each of the anteriormost anal rays. Thus, Böhlke's characters will not separate his two genera *Mimagoniates* and *Glandulocauda*, as the type species of the former genus lacks the characters supposedly diagnostic of it.

The caudal fin was seen to be split either to its base or to the midcaudal glandular area in *Phenacobrycon*, *Coelurichthys*, *Pseudocorynopoma*, *Landonia*, *Argopleura*, *Acrobrycon* (some specimens only), *Gephyrocharax chocoensis*, *Corynopoma*, and the Xenurobryconini among the genera examined.

Böhlke (1958a, p. 39) stated that "caudal hooks occur in relatively few Characid species." In view of the behavioral implications of hooks on fin rays (see Discussion), the fins of all genera except the xenurobrycones were examined. Males of all genera have variably placed hooks on the anal fin. Hooks are present on both pelvic and caudal fins of *Corynopoma*, *Gephyrocharax*, *C. microlepis*, the smaller specimen of *Compsura gorgonae*, *Acrobrycon* (only in this genus were hooks found on the dorsal fin), and according to Böhlke (1958a), *Hysteronotus*. They are present on the pelvics but not on the caudal in *Landonia*, *Phenacobrycon*, *Bryconamericus*, *Knodus*, *Hemibrycon*, *M. barberi*, *C. tenuis*, *Glandulocauda*, and *Planaltina*. They were seen on neither fin in *Pseudocorynopoma*, the larger specimen labelled *Compsura gorgonae*, *A. chocoensis*, and *Paragoniates*.

Finally, because Böhlke (1954) cited the apparently independently developed caudal scalation in *Compsura* as support for his thesis of polyphyletic origin for the Glandulocaudini, it is necessary to say a few words about the material of the present study labelled *Compsura gorgonae*. Two males, measuring 23 and 28 mm., were cleared, the latter unfortunately before being closely examined. The general appearance, color pattern, and anal fin counts of the two were identical, and both had modified caudal scales. One, the larger, lacked ventral and caudal hooks, but instead possessed a row of sharp, curved interhemal spines projecting from between the scales along the ventral edge of the caudal peduncle. These are characteristic of males of the genus *Cheirodon*, from which Eigenmann (1915) considers *Compsura* to be derived. The smaller (fig. 3, *a*) lacked these, and had hooks on both pelvic and caudal fins; in addition the caudal of this specimen was split to near its base and there was a slight suggestion of a glandular area between rays 10 and 11. In neither specimen was there a pouch under the modified scales. There are three "cheirodontin" species with modified caudal scales illustrated in the literature examined, *Odontostilbe hastata* Eigenmann (1915, fig. 33 and pl. XVI, fig. 1), *Compsura heterura* Eigenmann (1915, fig. 19 and pl. X, fig. 1), and *C. gorgonae* (Evermann and Goldsborough, 1909, figs. 1 and 3). None has the strong interhemal spines typical of *Cheirodon*. Very interestingly, *Compsura* and *Odontostilbe hastata* are found only in Panama and transandean Colombia, and in tributaries of the Rio São Francisco and the adjacent Rio Itapacuru 4,000 kilometers to the east (Eigenmann, 1915; Evermann and Goldsborough, 1909).

The results of this investigation are summarized in table 1. Their discussion will be postponed until after the behavioral and distributional evidence has been presented.

ECOLOGY AND DISTRIBUTION

Behavioral evolution in a group cannot be considered in isolation from the rest of its biology. Inferences as to probable function and evolution of behavioral mechanisms must be consistent with what is known of the ecology and distribution of the animals being studied; such considerations are included in the following paragraphs.

Habitat, Predation, and Food

Corynopoma will be considered first. This genus has been reported from Trinidad (see Boeseman, 1960, for extensive references and synonymy), the region around Villavicencio in the Colombian Llanos (Eigenmann, 1914), and recently from an affluent of the Rio Chirgua in Venezuela (Ramirez E., 1960). At the present time, *Corynopoma* appears to be a monotypic genus restricted to the Orinoco system, of which Trinidad may be considered a part (Price, 1955).

Collections made by me in the region around Villavicencio failed to reveal *Corynopoma*, although it was being particularly looked for, and although another Glandulocaudine, probably *Gephyrocharax*, was present in abundance. This and the dearth of reports of *Corynopoma* from mainland South America indicate that it is there either a rare species or one with a spotty distribution.

In Trinidad on the other hand, *Corynopoma* is one of the three most common species. As part of a general study of the freshwater fishes of Trinidad, a series of 284 collections from all accessible parts of the island was analysed. There were few consistent differences between collections on different dates, except that those collections which were obtained by means of electric shock (in 1958) generally contained a higher number of species. Relative abundance was only rarely recorded, and then only for commercially valuable species.

TABLE 1

PRESENCE OF SELECTED "GLANDULOCAUDINE" CHARACTERS

(+ = character present; (+) = character reduced; − = character absent; ? = character uncertain.)

GENERA	Midcaudal Glandular Area	Caudal Split at Least To Glandular Area	Pouch on Lower Caudal Lobe	Scales Extending onto Central Caudal Rays	Glandular Area on Lower Caudal Fulcra	Tubular Caudal Scale	Tubular Caudal Rays	Greatly Enlarged, Sculptured Caudal Scale	Hooks on Caudal Rays	Dorsal Origin Above or Behind Anal Origin	Markedly Oblique Gape
Glandulocaudini											
Corynopoma.....	+	+	+	+	+	−	−	−	+	+	+
Gephyrocharax...	+	+[a]	+	−	+	−	−	−	+	+	+
Pterobrycon......	?	−	+	−	?	−	−	−	?	+	+
Argopleura......	+	+	+	+	+	−	−	+	−	+[a]	+
Acrobrycon......	+	+[a]	+	+	+	−	−	−	+	−[b]	−
Planaltina.......	+	−	+	+	−	−	−	−	−	−	+
Phenacobrycon...	+	+	+	−	−	−	−	−	−	−	+
Hysteronotus.....	?	+[a]	+	+	?	+	−	−	+	+	+
Landonia........	+	+	−	+	−	+	−	−	−	−	+
Pseudo-											
corynopoma....	+	+	−	−	+	−	−	−	+	+
Coelurichthys....	+	+	(+)	+	+	−	+	−	+	+	+
Glandulocauda...	+	−	−	+	+	−	−	−	−	+[a]	+
Diapoma........	?	?	?	+	?	?	?	−	?	−	+
Xenurobryconini											
Xenurobrycon....	+	+	+	+	?	−	−	+	?	+	+
Tyttocharax......	+	+	+	+	?	−	−	+	?	+	+
Other Characidae											
Compsura.......	(+)[a]	+[a]	−	+[a]	−	−	−	−	+[a]	−[b]	−
Paragoniates.....	−	−	−	−	−	−	−	−	−	+	+
Bryconamericus..	−	−	−	−	−	−	−	−	−	−	−
Hemibrycon......	−	−	−	−	−	−	−	−	−	−	−
Knodus..........	−	−	−	−	−	−	−	−	−	−	−

[a] In some specimens or species.
[b] Only in these two genera is the dorsal fin considerably forward of the anal fin consistently.

In Trinidad, *Lebistes reticulatus*, the guppy, is by far the most widespread species. If it does not occur naturally in certain areas it has been introduced as a mosquito control measure (Guppy, 1934). The next most abundant are *Corynopoma* (in 178 out of 284 collections) and *Astyanax bimaculatus*, a characid species widespread on the continent (in 176 collections).

Trinidad probably contains a wider range of habitats than most mainland areas of similar size, and nearly all of these are present in the very accessible area in the Northwest which is drained by the Caroni, the largest stream on the island. The 99 samples taken from the Caroni drainage were given special attention. *Coryno-*

poma and *Astyanax* were each present in 54 samples of the 99; chi-square tests (Dice, 1952, p. 317–318) showed *Corynopoma* to be significantly associated with *A. bimaculatus* ($p<0.01$) and three other species: another characid, *Hemigrammus unilineatus* ($p<0.01$), a callichthyid catfish, *Corydoras aeneus* ($p<0.01$) and a cichlid, *Cichlasoma bimaculatum* ($p<0.02$). All four are cosmopolitan species ranging over much of tropical South America. That the sampling was not biased in terms of depth of water is indicated by the fact that *Corydoras* is a bottom dweller, *Cichlasoma* tends to stay near the bottom, *Astyanax* ranges freely over all depths, and *Corynopoma* usually remains within a foot or so of the surface. Also, if the observed association were due to the sort of bias which the occasional use of electric fishing would introduce, one would expect collections containing rarer species to contain higher numbers of species than the average. This was the case only for the four rarest species of the twenty-three which were present in more than 4 percent of the samples. For species present in more than 5 percent of the samples, no such correlation existed.

The observed associations were rather due to common exclusion from particular habitats. *Corynopoma*, *Cichlasoma*, *Hemigrammus*, *Corydoras*, many of the less common species and, to a lesser degree, *Astyanax*, seem to be excluded from those valleys having an average gradient of more than about 50 feet per mile. At the other extreme, *Corynopoma*, *Hemigrammus*, *Corydoras*, many of the other species and to a lesser extent *Cichlasoma* and *Astyanax* appear excluded from those habitats likely to become brackish in the dry season.

Within these limits, however, *Corynopoma* was very nearly ubiquitous, and although there were indications that it was less abundant in clear as opposed to turbid or coffee-colored water, its occurrence or non-occurrence could not be correlated further with particular habitats. It was abundant in wide, turbid, swiftly moving streams, ditches in rice fields, irrigation ditches in reclaimed swampland, ponds in pastures thick with algal bloom, isolated pools in dried-up creeks, deep pools in streams through cane fields, and the wide, slowly moving Caroni River several miles from its entry into the brackish Caroni Swamp. The ubiquity of *Corynopoma* may be illustrated by the following. Of the 142 collections from streams and ponds in the drainage of the Central Range, covering nearly all kinds of habitat on the island except those associated with greater-than-one-percent gradients, 123 or 87 percent contained *Corynopoma;* the corresponding figure for *Astyanax* was 113, or 80 percent.

Juveniles were not particularly correlated with specific habitats, although they appeared more common in irrigation ditches and in some of the large, turbid streams. The subject of seasonality of their appearance will be taken up in a later paragraph.

The turbidity of the water in most *Corynopoma* habitats precluded study of the genus in its natural environment; however, some observations and a few experiments carried out in tanks and pools at Bamboo Grove may be briefly reported. In deep tanks containing various combinations of species, *Lebistes*, *Corynopoma*, and *Hemibrycon* sp. aggregated together or separately near the surface. The *Hemibrycon* continually chased and nipped at the others, which were of similar size. *Astyanax* ranged through all levels; other characids schooled or aggregated in midwater or remained, aggregated or isolated, near the bottom. Predators, except for *Polycentrus*, usually remained near the bottom.

Two large tanks, each containing clear water and no vegetation, were stocked with four each of the following characids, all of approximately the same size: *Corynopoma, Hemibrycon* sp., and *Pristella riddlei* (or possibly *Aphyocharax axelrodi;* see Boeseman, 1960). In one tank were several *Polycentrus schomburgkii;* in the other were one large predaceous characid, *Roeboides dayi,* a young *Crenicichla alta,* and a large goby, *Dormitator maculatus.* All of these predators attacked all three prey species, but they, and especially *Polycentrus,* caught *Corynopoma* with far greater facility. After one hour the *Polycentrus* tank contained four *Pristella,* three *Hemibrycon,* but only one *Corynopoma;* after three hours only the four *Pristella* and two of the *Hemibrycon* remained. Results in the other tank were similar but took longer to achieve. It is interesting that *Hemibrycon* and *Pristella* have contrasting spots. In *Pristella* these are on the dorsal and anal fins rather than on the caudal peduncle, and *Pristella* escapes with an up or down movement rather than with the more usual forward jump.

It is possible that this apparent vulnerability of *Corynopoma* to predation may be involved in its apparently restricted clear-water distribution. A chi-square test for significance of association, using the 247 samples from habitats in which both *Polycentrus* and *Corynopoma* might be expected to have occurred, showed that samples containing both species were significantly less frequent than would be expected to occur by chance ($p < 0.02$). However, an undetected habitat difference might have produced these results (*Polycentrus* was indeed more frequently encountered in clear, stagnant water).

A word may be said here about feeding habits. Guppy (1934) gave some attention to *Corynopoma* as an important larvicide (to my knowledge it is not intentionally so used in Trinidad). Aquarium observations indicate that it also includes in its diet small crustacea and floating and perhaps low-flying insects. Release of a cloud of *Drosophila* above a tank unleashed a burst of vigorous breaking of the surface. It was not clear whether the insects were actually caught on the wing or whether droplets of water produced by the splashing first brought them to the surface, where they were subsequently caught.

It is certain that in Trinidad at least, absence of mosquito larvae does not restrict its distribution. Samples from sewage outfalls, where mosquitoes were rare, contained *Corynopoma* in abundance; they were also common in acid, coffee-colored water, in which mosquito larvae do not do well (Hesse, Allee and Schmidt, 1951, p. 440).

Water temperature would not seem to be a serious limiting factor for *Corynopoma,* at least in Trinidad. Water temperature probably does not drop below 20°C, at any time, except possibly in the mountain streams in the northern range. During August and September, 1961, the lowest temperature recorded was 23°C. The highest recorded was 33°C., in a ditch on the Kaltoo Trace in Nariva Swamp, and on both occasions *Corynopoma* was present. Undoubtedly the temperature rises higher in temporary ponds, and there temperature or an accompanying oxygen deficit may limit distribution of the species.

Harald Schultz (1959 and pers. comm.) has given the following information on the southeastern Brazilian species. *Coelurichthys tenuis* (= *M. barberi* of H. Schultz) probably ranges from São Paulo (24°S.) to Santa Catherina (28°S.). It is found only in the quiet stretches of small swift coastal streams in which the water is clear, deep brown, and acid. These places are usually overhung with grasses and contain clumps of *Utricularia*. *C. tenuis* schools at a depth of 10–12 inches in the open water.

The smallest fry are never found in this situation, but rather in tiny tributary streams, sometimes only 10 inches wide and drying out in the dry season; grown specimens have never been observed there.

C. (= *M.*) *microlepis* occurs from above Rio de Janeiro (22°S.) to Santa Catherina (28°S.).

Farther south its place is taken by *Glandulocauda inequalis* (unless of course the scientists are correct in assuming that both species are the same). In contrast to *M. barberi* [*C. tenuis* of the present paper], *M. microlepis* occurs in fairly large streams. *M. microlepis* and *G. inequalis* are found in stretches of clear, colorless water. In some places one can find both species together, but never in equal numbers. Like all *Mimagoniates*, *microlepis* prefers calm spots in rapidly flowing streams, particularly in the shade of branches or the overhanging grass of the shore. Here we can find them in schools of several hundred.

When a brook was cleaned out by cutting the grass on the shore and removing the branches and other foreign matter the water became clouded. The otherwise numerous *M. microlepis* disappeared for quite a time. (H. Schultz, 1959, pp. 52–53).

In this connection H. Schultz points out that at least in the brown-water streams, the annual floods do not cause an appreciable change in turbidity or other properties of the water. The opposite is true in both the Colombian Llanos and especially in Trinidad (pers. obs.).

The colorless waters inhabited by *C. microlepis* are approximately neutral in pH; in the more southerly habitat of *G. inequalis* and *Ps. doriae* they tend toward acidity (pH 6.8–6.6 at Guahyba Bay). Here *Utricularia* is replaced by *Myriophyllum* and *Nitella*, but otherwise conditions are similar. A. V. Schultz (1962, p. 9) writes of *Ps. doriae:* "The fact that I caught—and still catch—the greatest number of young specimens in a swamp near a creek seems to indicate that in flood time the fish like to follow the overflow of waters, spawning where the current is not as strong as in the brooks and rivers themselves, and where food animals seem to be more plentiful. But this is not an absolute rule, for I have also caught baby Dragonfins in brooks and creeks..."

Water temperatures near Santos range from 21–25°C.; near Porto Alegre they range from 15–26°C. Occasionally more extreme temperatures may be found at both localities.

Breeding Season

The question of a breeding season may now be discussed. Under aquarium conditions courtship may be seen at all times of the year in *Corynopoma*, and this species exhibits consistent seasonal differences in neither intensity of courtship nor frequency of spawning. Courtship was seen in *Coelurichthys* and *Glandulocauda* with equal intensity and frequency in all months. With *Ps. heterandria*, sporadic courtship in both male and female was seen beginning in December. In *Ps. doriae*, sporadic courtship by the male occurred every month, but only became intensive in mid-February. Female responses began in February.

Field observations are somewhat more conclusive. Price (pers. comm.) states that the peak in the number of young *Corynopoma* found occurs during the rainy season. H. Schultz (1959) writes that in *Coelurichthys* and *Glandulocauda* spawning probably takes place shortly after the beginning of the annual floods. He emphasizes, however, that in the region of Santos these are accompanied by a rise in temperature, whereas in Rio Grande Do Sul "The seasons are reversed. Therefore they have a hot, very dry *summer* (December to March) and a cool, rainy winter with floods in September." In *Pseudocorynopoma* also spawning occurs in September, "The first month of spring here in the Southern Hemisphere" (A. V. Schultz, 1962, p. 9).

DISTRIBUTION

More generally, evidence on the biogeography of the Glandulocaudini is fragmentary and contradictory. In 1929, Eigenmann and Myers remarked on the fact that the then-known Glandulocaudines were confined to the northwestern and southeastern extremes of tropical South America. Since then the gap has been closed, however, and even in 1929 Myers (Eigenmann and Myers, 1929) had described a species of *Gephyrocharax* (*G. major*) from a tributary of the Madeira in Bolivia, and *Acrobrycon ipanquianus* was known from the same region.

So far as is known, nearly all glandulocaudine genera have restricted distributions. *Hysteronotus,* an exception, is known only from a small area in Peru and from the São Francisco system, 4,000 kilometers to the east (Böhlke, 1958a).[3] *Planatina myersi,* known only from a single specimen from near the headwaters of the São Francisco, is most closely related to *Argopleura* from transandean Colombia, *Phenacobrycon* from Peru, and *Acrobrycon* from Bolivia (Böhlke, 1954, and pers. obs.). Thus these two disjunct distributions (and that of *Compsura,* see preceding section) are nearly parallel. The other exception is *Gephyrocharax,* the only genus with a cosmopolitan distribution. *G. major* from the Rio Beni in the Amazon basin is scarcely specifically distinct from *G. chocoensis* from the Rios Atrato and San Juan in transandean Colombia (Myers, in Eigenmann and Myers, 1929), and the specimens labelled *G. chocoensis* examined in the present study were from Bolivia.

There is evidence for geographical replacement among the southern Brazilian species (see above) and in the *Gephyrocharax* of Colombia, where each of the major intercordilleran rivers has its own quite distinctive species (Eigenmann and Myers, 1929). A similar situation appears to exist among *Argopleura* of the same region (Eigenmann, 1927).

The apparent ecological replacement in *Coelurichthys* has been outlined above, and it is possible that a similar situation exists with *Corynopoma* and other Glandulocaudines on the continent. However, in Trinidad *Corynopoma* behaves as a cosmopolitan species, not as one fortuitously occupying a vacant niche. Factors other than the absence of competing species, such factors as the relative absence of pelagic predators and the changes wrought by intensive cultivation in Trinidad, may contribute to the ubiquity of *Corynopoma* there.

---

[3] Recently, a collection of *Hysteronotus* from Venezuela has come to light (Weitzman, pers. comm.).

NON-SOCIAL BEHAVIOR

Observations on the behavior of isolated fish must provide a baseline from which to assess the animals' performance in social situations; while in addition, non-social behavior may be expected to provide an evolutionary source for movements with social significance. Spawning behavior has been included in this category, for in these internally fertilizing fishes the female may spawn in the absence of other fish.

LOCOMOTION AND ORIENTATION

The behavior of *Corynopoma* will be described first. Both males and females maintained themselves in practically continuous motion, the basic component of which will be called the Beat: a rapid movement to one side and back of the head, followed rapidly by the rest of the body and the caudal fin as a single wave, and accompanied by a sharp beat or closing of the pectoral fins and a moderate, apparently passive closing of the median fins. A Beat would last about one-fifth of a second or less, and was often followed by a Glide, during which the pectorals and partially closed median fins returned to their former positions, generally at angles of 45–60° to the body. The Glide lasted for as long as two seconds.

Rapid locomotion appeared to be largely a matter of eliminating the Glide and increasing the strength of the pectoral beats; rarely did two Beats overlap so that there was more than one wave traversing the fish at the same time.

The alternate Beats resulted in a slightly zigzag locomotion when seen from above. Braking was done by a forward sweep of the pectorals and was often accompanied by a near-maximal erection of the median fins. Turning in a vertical plane was accomplished by an up- or down-twitching of the pectorals. Turning in a horizontal plane was generally a combination of unilateral pectoral beating, braking or twitching, often with several Beats.

Often the fish was seen to swim up and down the side of the tank with a shimmying movement composed of alternate, rhythmical, overlapping Beats. The angle of the body with the glass decreased as the up-and-down speed increased; the median fins were often nearly maximally erected.

During horizontal movement the fish usually were oriented about as in figure 1, *a*. When maintaining a horizontal position, however, a fish seemed almost always to have to compensate for a slight tendency of either its head or its whole body to rise or to fall. Occasionally while the fish maintained a slow forward locomotion or while it remained stationary, the upper caudal lobe vibrated or twitched, pivoting about the top ray.

Locomotion and orientation in the other Glandulocaudines were in general similar. In *Pseudocorynopoma*, *Coelurichthys*, and *Glandulocauda* the pectorals appeared to provide very little driving force during the Beat. Resting *Pseudocorynopoma* of both species usually oriented head raised, with the longitudinal body axis of normal locomotion at a 10–15° angle with the horizontal. *Pseudocorynopoma* usually moved more deliberately than did the other species, almost sedately in fact. Upon being frightened, *Ps. heterandria* occasionally leaped from the water, often to land in adjacent tanks. This was never seen in *Ps. doriae*. The form of these leaps was not studied, but from the configurations of the aquaria, water levels and cover glasses, it could be deduced that at times the leaps were nearly vertical.

In *Glandulocauda,* fusion of Beats was relatively more frequent than in the other species, which in this respect resembled *Corynopoma* closely. The two *Coelurichthys* species and *Glandulocauda* were more variable in their resting orientation than was *Corynopoma*. In these three species Glides were on the average shorter in duration than in *Corynopoma;* in *Pseudocorynopoma* they were longer.

Cursory observations of locomotory behavior were made of the following characids: *Aphyocharax rubripinnis, Prionobrama* sp., *Hemigrammus unilineatus, Bryconamericus* sp., *Knobus* sp., *Hemibrycon dentatus, Hemibrycon* sp., *Moenkhausia oligolepis, Astyanax bimaculatus,* and *Triportheus* sp. Locomotion and stationary movements in all of these species but the first two were similar, and closely resembled those described above for the Glandulocaudines. *Aphyocharax* and *Prionobrama* were exceptional in that in stationary fish, the dorsal fin and the upper lobe of the caudal were in nearly continuous rhythmic motion, often the pectorals were seen to beat alternately and rhythmically (Wickler, 1960), and the fish was often seen to back up. Rhythmic pectoral movements were never seen and backing was seen only rarely in the other species. In their locomotion and orientation, the subadult *Triportheus* most closely resembled *Pseudocorynopoma* of all the species examined.

Feeding Movements

In all Glandulocaudines studied, feeding on suspended particles was largely accomplished by a quick up-and-down movement; the fish moves down and forward, its mouth opens, the fish moves up and forward, its mouth closes. If the fish was hungry, and there was an abundance of suspended food, seven or eight complete movements in all directions would be accomplished in five seconds. In this case the pectorals seemed to provide most of the locomotive power. In *Pseudocorynopoma* these high frequency bouts were less frequently seen, and a greater proportion of approaches to the food were from below than in the other species.

Feeding from the surface appeared to take much the same form; feeding from the bottom or from the surface of the leaves was much more deliberate, with a more obvious searching or inspecting component. The body was held at a 50–90° angle with the substrate, and backing sometimes followed the forward movement. Feeding from the bottom was very rare in all Glandulocaudines, and rarest of all in *Pseudocorynopoma*. Backing movements were confined to this type of feeding.

Triportheus resembled the Glandulocaudines in feeding on suspended matter with an up-and-down movement. Feeding was similar in *Hemigrammus, Bryconamericus, Knodus, Hemibrycon, Moenkhausia,* and *Astyanax* except that the up-and-down component was less pronounced and they were observed to feed from the substrate much more often. *Aphyocharax* and *Prionobrama* tended to hover in front of a particle of food, and then take it with a sudden forward movement. They spent much time browsing on "Aufwuchs" (periphyton and associated animals) in this fashion.

OTHER NON-SOCIAL BEHAVIOR

In Glandulocaudines it was difficult to distinguish movements serving to relieve discomfort ("Comfort Movements," Baerends and Baerends, 1950), and those which appear to be stretching movements precedent to some unspecified activity ("Intentionsbewegungen" in the sense of Kortlandt, 1940). Tentatively, all movements shown independently of a social context (but in social situations as well, as will be seen), which do not obviously subserve alimentation, respiration or locomotion, will be designated "Comfort Movements."

The majority of the comfort movements exhibited by the characids studied appear to be interrelated and are for the most part quite unexceptional.

Chewing, a rapid up-and-down movement of the lower jaw, may be considered first. Most of the time the fish apparently had nothing in its mouth, but often it would dash to the surface and return chewing a piece of duckweed or other debris. This will be called Nibbling. The fish was never observed to swallow the particle, but only to worry continually at it, chewing, spitting it out, chewing, etc., often for as long as 90 seconds. It is included under the heading of comfort movements only with hesitation, for there is always the possibility that the fish nibbled duckweed for its juices or attached invertebrate life. More often, upon dashing to the surface the fish returned with only a bubble of air, or with nothing. This has been termed Nipping Surface (NS) and will be discussed in a later section. Both Nibbling and Nipping Surface were included as comfort movements because of the lack of a clear-cut relation with feeding behavior.

Chewing and Nibbling were often accompanied by Fin-flicking, a rapid raising and lowering of the median fins, and in *Corynopoma* males by Paddle-flicking, an outward jerk of one or both opercular appendages.

More spectacular were Gaping, Fin-raising and, in *Corynopoma* males, Paddle-raising, often seen together, as in figure 5, c. These actions resembled Chewing, Fin-flicking and Paddle-flicking, but were more extreme and prolonged. In males of *Pseudocorynopoma* and *Corynopoma*, in Fin-raising the dorsal and anal fins were supramaximally erected (fig. 5, c). The complex of Gaping and Fin-raising, when they occur together, will be called Yawning.

The median-fin movements were less extreme in females of all species, and seemed to be balancing movements when they accompanied Chewing or Gaping. In both sexes, all raisings of the median fins were accompanied by (usually downward) flicks of the pectoral fins (see fig. 5, c)

In male *Corynopoma* Yawning was accompanied by a downward rotation of the caudal peduncle as in figure 5, c. In Yawning, male *Pseudocorynopoma* and *C. tenuis* often threw themselves into a sharp sigmoid flexure, returning to a normal posture by an exaggerated Beat.

Rarely, males were seen Chafing their sides against beds of *Fontinalis;* more rarely, females would Chafe their bellies against the fine white bottom sand.

The observed Comfort Movements may be summarized as in table 2 which, read across, emphasizes the identity of effectors, and read up and down, emphasizes contiguity in time.

Spawning

Spawning was witnessed five times in *C. riisei*, twice in *C. microlepis*, and once each in *C. tenuis* and *G. inequalis*. In all cases it began before 9 A.M. and was over by 1 P.M. Pacific Standard Time. For all species the preferred substrate was giant *Sagittaria*; however, plastic filter siphons and the glass aquarium walls were also used by *Corynopoma* and to a lesser extent by *Glandulocauda*. Materials which were not used were fine-leaved plants (*Heteranthera, Fontinalis*), rocks, and gravel.

In *Corynopoma* no preference could be detected between transparent aquarium walls and walls backed with green or black paper. With this species detached *Sagittaria* leaves were spawned upon only if more than several inches long, indicating that the resistance to push provided by the substrate may be important.

TABLE 2
Comfort Movements

Mouth	Chewing Nibbling	Gaping	
Median Fins	Fin-flicking	Fin-raising	
Opercle (*Corynopoma*)	Paddle-flicking	Paddle-raising	
Belly and Sides			Chafing

Corynopoma females were observed first to approach the substrate perpendicularly, and then to mouth or bite at it. Then the female moved upward, rotating so as to place the short ovipositor near the point mouthed. The ventral surface was then pressed to the substrate and as the female moved along it two to eight eggs were expelled one by one. Occasionally this forward movement was not continuous but rather in discrete jerks. This was most often seen when the substrate was a *Sagittaria* leaf; with each jerk sufficient force was applied to move the leaf slightly. The result of each such spawning act was a string of eggs generally spaced less than a millimeter apart.

During spawning female *C. microlepis* were observed to bump a prospective substrate with the nape rather than with the mouth. This was seen only a few times with *C. tenuis*. In both species it was usually the undersides of the leaves which were selected. The female would hover with the leaf to one side of her head, then rapidly move forward and rotate so that her ventral surface struck the leaf a glancing blow. Usually only one egg was laid at a time. In *C. tenuis* the female rubbed, rather than struck, the leaf with her ventral surface; the usual result was still the deposition of only one egg.

The female *Glandulocauda* that was observed followed in general the procedure outlined for *Coelurichthys*, but testing of the substrate with the nape was not noted. The ventral surface was pressed to the glass or to the underside of a leaf with a little quiver, not seen in the other species. Only one out of every four or five of these resulted in the deposition of eggs. One, two or three eggs were deposited at a time.

Innes (1948, page 150) wrote with regard to *Corynopoma riisei:* "The female very soon after spawning begins to move the eggs from one point to another, taking them in her mouth and sticking them on a new spot on a leaf. She constantly swims about them and guards them, refusing all food until several days after they have hatched." In the present study, both aquarium specimens and wild-caught *Corynopoma* were observed to spawn and their subsequent behavior was recorded. Although females of all four species were somewhat more aggressive before, during, and after spawning than at other times, they quickly returned to normal, at no time refused food, and in no case was defense or moving of eggs recorded. All species, however, were seen to favor certain sites over other apparently just as suitable ones. In particular, one female *C. microlepis* attempted to deposit over one-half of her eggs on two *Sagittaria* leaves; toward the end of the spawning one, two, or more eggs would be dislodged for each egg successfully laid, and she never failed to catch and swallow these before they reached the bottom.

ELEMENTS OF SOCIAL BEHAVIOR

In the Glandulocaudines studied, there were few hard-and-fast distinctions between *actions* an animal would show with its own sex and with members of the opposite sex; rather, the *patterning* varied with the context. Accordingly, it has been thought best to include here all actions with a social context and to reserve attempts at correlating these actions for the following sections. Each action will here be classified and described according to the sex of its performer and according to the situations in which it is seen. In each category, those actions common to most species will be described first, and these will be followed by descriptions of actions peculiar to particular species. While it is recognized that in this procedure there lies the danger of creating false homologies by fiat, this risk is deemed preferable to the needless repetition which would be involved in devoting separate accounts to each species.

The only functional assumption in the following classification is that all actions therein described are either communicatory or ancillary to communication. Nevertheless, it is important to realize that this classification is a hybrid one, resting upon both contextual and morphological criteria.

Actions Performed by Both Sexes

Approaching (A).[4]—This term will be used to refer both to slow approach to one individual by another and to the maintenance by the latter of a constant distance behind the former, as in practice the two types were often indistinguishable. Approaching appeared in many contexts and in all species, often as an orienting action preparatory to more specific activity. It was also the most frequent first response to a newcomer. In the more slowly moving *Pseudocorynopoma* it appeared in place of Chasing (see below) even in rather violent aggressive encounters.

---

[4] The letters in parentheses will be used in the following sections to refer to the corresponding activities.

Chasing (C).—The performer followed another fish rapidly. Whereas Approaching did not usually result in overtaking, Chasing always resulted in either overtaking or in violent contact, unless the pursued fish Moved Away rapidly (see below). Whereas Approaching often took the form of a single Beat and Glide or else was maintained by primarily pectoral movements, Chasing was mostly accomplished by a series of strong Beats, either individually distinguishable or fused together.

In *Corynopoma*, Chasing by the male often served to orient him ahead of and parallel with the female. Various activities which are clearly separable from Chasing in the other species studied merged imperceptibly with it in *Corynopoma*, and in the timed records it was not found practicable to separate the variously intergrading actions. These will be noted, however, in the proper places below.

In *Coelurichthys* and *Glandulocauda*, the male would very often follow a rapidly swimming female about the tank, reflecting her movements with remarkable precision. In *C. tenuis* and *Glandulocauda* these bouts of Chasing would often be unbroken for as much as two minutes.

In *Pseudocorynopoma*, Chasing, as defined, was nearly restricted to aggressive encounters, and was nearly always followed by Biting.

Biting (B).—Usually Biting was directed at the caudal fin or at the flanks; Biting may or may not have made contact. It was often impossible to tell whether it had, and in cases of doubt subsequent dashing away by the victim was used as the criterion of whether Biting had occurred. In *Pseudocorynopoma* in particular, the bites were nearly always directed at the flanks, and the flash of scales suspended in the water was an invariable sign that the Biting had connected. Biting was nearly always preceded by a short Chase.

Moving Away (M), *slowly* (Ms) or *rapidly* (Mr).—Moving Away occurred either as a response to an Approaching or Chasing animal or without apparent cognizance of a following animal, or, cognizant or not, relative to a non-following animal. These will all be designated as Moving Away slowly. Thus, any form of locomotion suitable to slow movement may be employed. Moving Away rapidly, on the contrary, always occurred after some action by another animal, usually of the same species, and apparently as a reaction to it. Moreover, it was always accomplished by strong Beats, rarely fused together.

Dropping (Dr).—In this, a fish sank to the bottom of the tank and remained there, often with head higher than tail and usually moving its pectorals to compensate for a strong tendency to sink still further. This was especially common among fish which had been Chased and Bitten considerably, but also occurred among fish recently handled roughly, placed in new surroundings, or subjected to a sudden increase in illumination. Often the slightest Approach by a Biting animal was sufficient to elicit Dropping.

Lateral Display (LD), *fins up* (LDfu) or *fins folded* (LDff).—Here, sufficient species differences existed to warrant considering the genera separately.

In both species of *Coelurichthys* the typical form of the display was a Glide in which the pectoral fins were held outward and apparently used as hydrofoils, the median fins were supramaximally erected, and the fish presented its broadside to its opponent. In *C. tenuis,* the branchiostegal apparatus was dropped forward and down, and the pelvic fins were spread. In both species the mouth was often open. This will be called Lateral Display, *fins up*. In the two *Coelurichthys* species, the apparent increase in size was most striking in the males, in which the expanded anal fin outline became continuous with that of the lower lobe of the caudal (fig. 6, *b*). In *Coelurichthys* four predominant orientations of the fish relative to one another during Lateral Display may be described:

1. *Circling.* Two fish circled, head-to-tail. The circle so described lay in either an horizontal or an oblique plane (fig. 6, *a*).

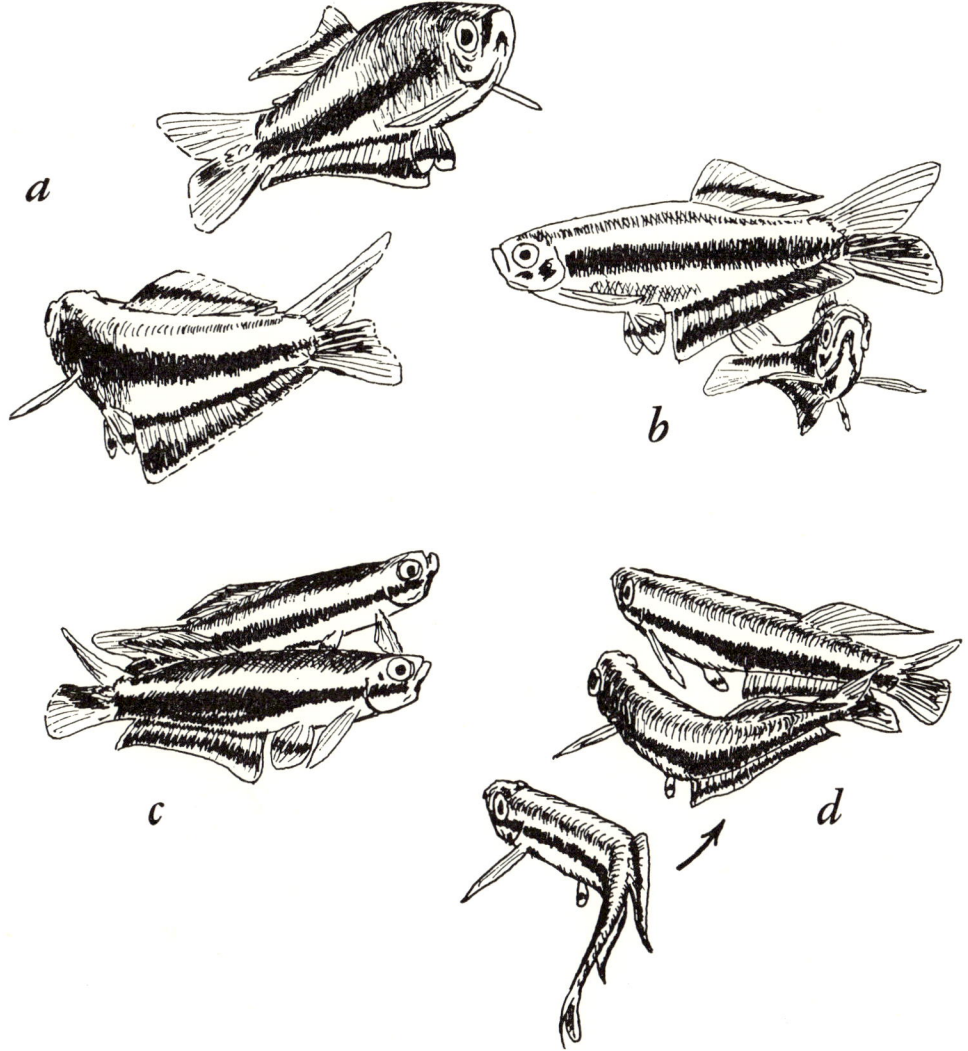

Fig. 6. Aggressive behavior, *Coelurichthys:* *a.* Lateral Display; two *C. microlepis* males, Circling. *b.* Lateral Display; male *C. microlepis*, Broadside to another animal. *c.* Lateral Display; two *C. tenuis* males, Head-To-Head. *d.* Tail-Beating; *C. tenuis* males, showing two successive positions in the Tail-Beat. (From photographs.)

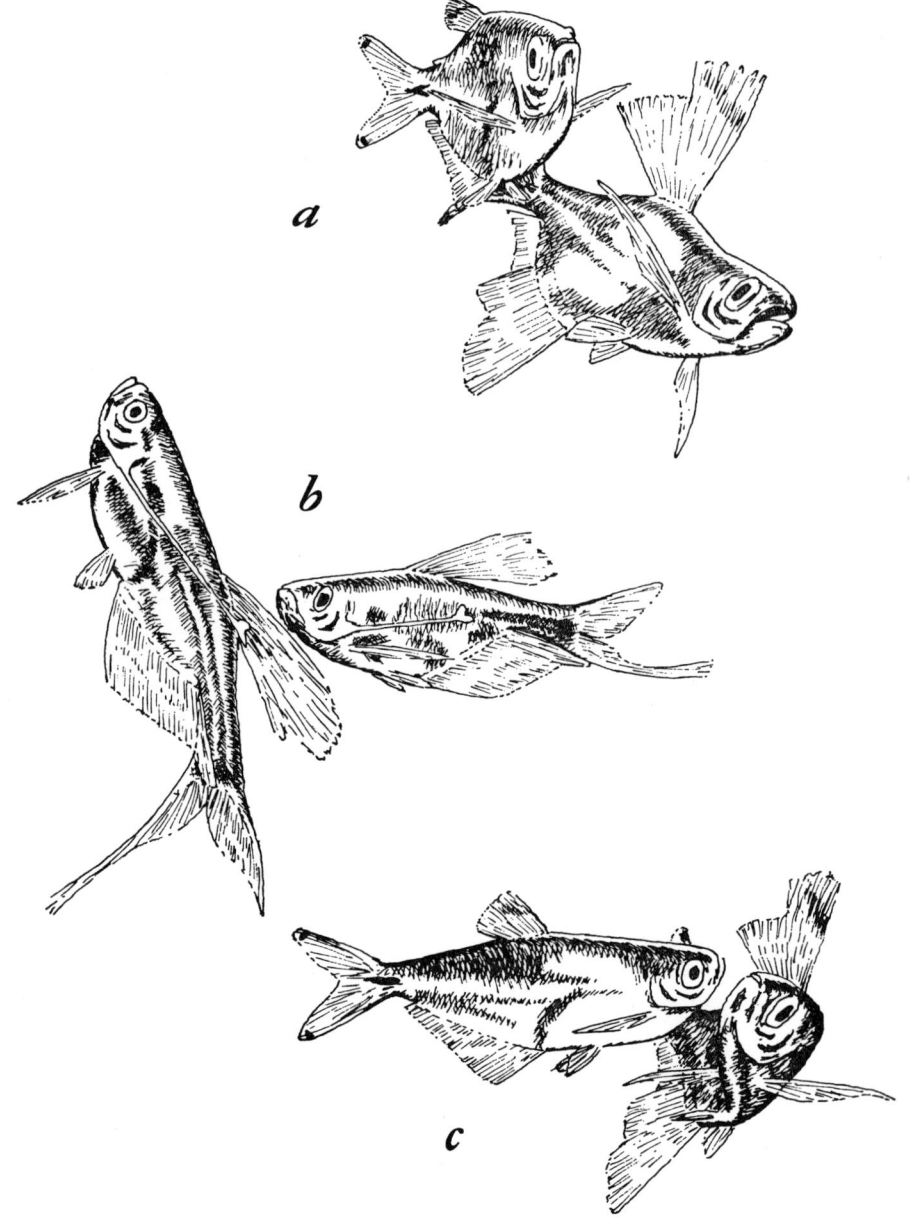

Fig. 7. *a.* Wobbling by a *Pseudocorynopoma doriae* male (right) under a female. (From photographs and sketches.) *b.* A male *Corynopoma*, Upright Posturing in front of another male. (From a photograph.) *c.* A male *Pseudocorynopoma doriae* passing in front of the female during Ovalling. (From photographs and sketches.)

2. *Broadside.* The displaying fish glided in front of the spectator, forming the bar of a T (fig. 6, *b*).

3. *Double Broadside.* Two fish swam toward and glided past one another. As they passed they usually rotated in a vertical plane, head upward, to an angle with the horizontal of about 25°.

4. *Head-to-Head.* Two fish swam toward one another, generally at an angle, then both turned (toward the vertex of the angle) and swam or glided side by side. (fig. 6, *c*).

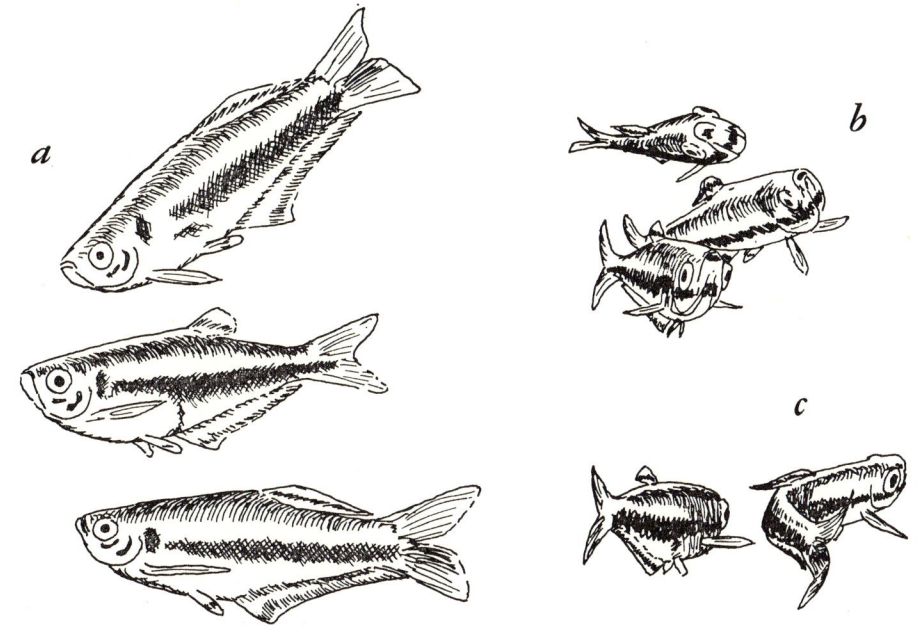

Fig. 8. Courtship in *Coelurichthys*: *a.* Hovering in *C. microlepis*, showing a male Hovering below and above the female (in the center). *b.* Hovering in *C. tenuis*, showing two orientations during Hovering, above and to one side of the female (lower animal). *c.* Dusting in *C. tenuis*, showing the position of the male (right) and female at the start of the tail-swing. (From photographs.)

All four orientations were seen to occur with the fins either raised or depressed. In *C. tenuis*, Head-to-Head orientation often led into mutual Pairing, below.

In both species of *Pseudocorynopoma* the anal fin was usually supramaximally unfolded, the dorsal fin held at an angle of 45° during Lateral Displays, and the anal fin was more often folded in the female. All four orientations were seen; often a Broadside would be interrupted by a Bite by the spectator.

In the case of *Corynopoma* Lateral Displays, which were in any event quite rare, were never seen to occur with the fins unfolded. Of the four *Coelurichthys* orientations, there were seen in *Corynopoma* only the Broadside and a modified version of it in which the displayer made an acute angle with the line of approach of a following animal. The modified version was usually begun with an exaggerated Beat of large amplitude and little propulsive power. It often led into the Upright Posture (see below).

Head-On orientation (HO).—This was seen only in *Coelurichthys*. Two animals hung suspended in the water, facing one another, at a distance of 2–3 cm. The body axes were usually inclined, head up, and pectoral movements compensated for an apparent tendency to drop. Often three animals would engage in this, radiating from a common center like spokes. This orientation would often persist for minutes on end.

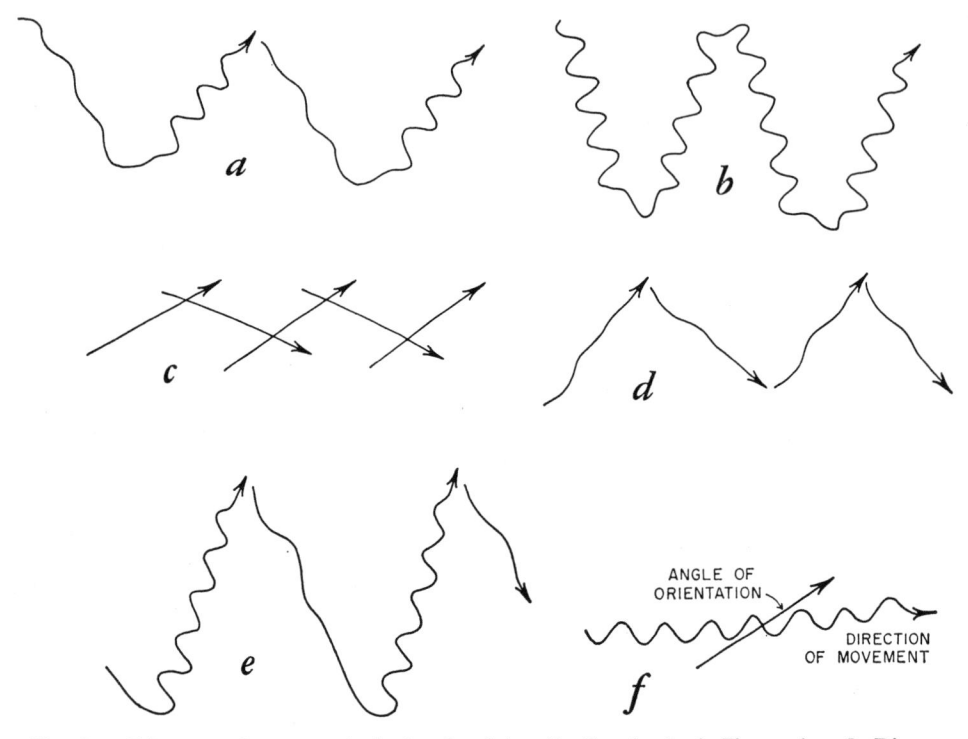

Fig. 9. *a*. Diagram of movements during low-intensity *C. microlepis* Zigzagging. *b*. Diagram of high-intensity *C. microlepis* Zigzagging. *c*. and *d*. Two forms of Zigzagging in *C. tenuis*. *e*. Zigzagging in *Ps. doriae*. *f*. Wobbling in *Ps. doriae*, showing the difference between the angles of orientation and the direction of movement. (Lateral undulations are here represented schematically as in the vertical plane. Further explanations in the text.)

Tail-Beating (TB).—This again was confined to *Coelurichthys* and was rare in *C. tenuis*. An animal would approach another, flexed into a V-shape, and using largely the pectorals for locomotory force (fig. 6, *d*). At a distance of about 1 cm., the tail would be thrown to the other side, so that the apex of the V, previously toward the recipient, now pointed away from it. This would often be followed by several exaggerated stationary Beats by one or both animals. When seen in *C. tenuis*, Tail-Beating often led into Lateral Display, Head-to-Head.

Wobbling (W).—This was confined to the two species of *Pseudocorynopoma*. Often while Zigzagging about the female (see below), a male would gradually or suddenly eliminate the downward component and, with lateral undulations at a frequency of two to four per second, pass back and forth and in circles beneath and alongside her. His head was tilted upward at an angle of 30–60° (fig 9, *d*), and he maintained the plane of his body perpendicular to a line passing between them (fig. 7, *a*). If she Followed him, he often increased the frequency of undulations; if he did so, she would often follow after him, also Wobbling. In Wobbling the median fins were unfolded and the pectorals were held straight out from the body and seemed to be used as hydrofoils.

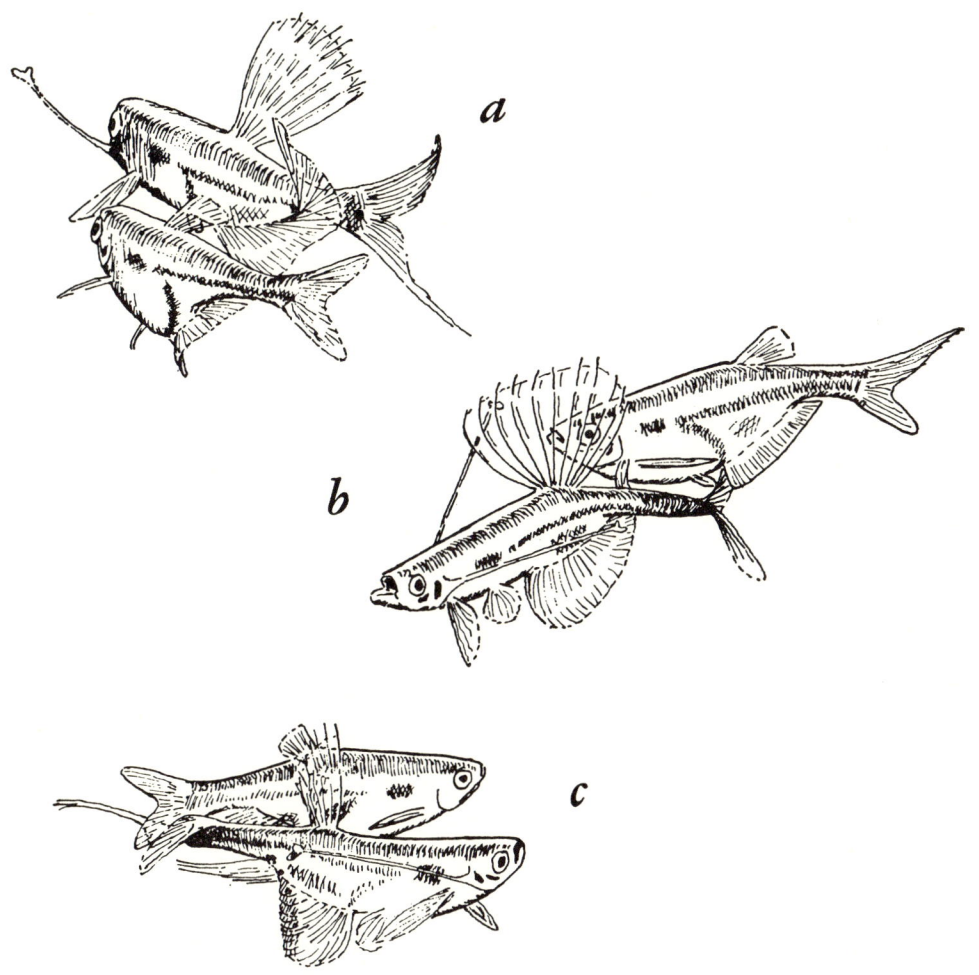

Fig. 10. Courtship in *Corynopoma*: *a*. Male Extending to a Following female. *b*. Male Shaking. *c*. Male (lower animal) Quivering. (From photographs. Explanation in text.)

ACTIONS PERFORMED ONLY OR PREDOMINANTLY BY MALES

Upright Posture (U).—This was seen only in *Corynopoma* and *Pseudocorynopoma*. In *Corynopoma* it consisted of an oriented Beat-Glide, accompanied by a downward pectoral thrust which both braked the glide and sent the performer into an oblique or vertical position with its head up. Usually the performer would finish the movement directly in front of the other fish and with its long axis at

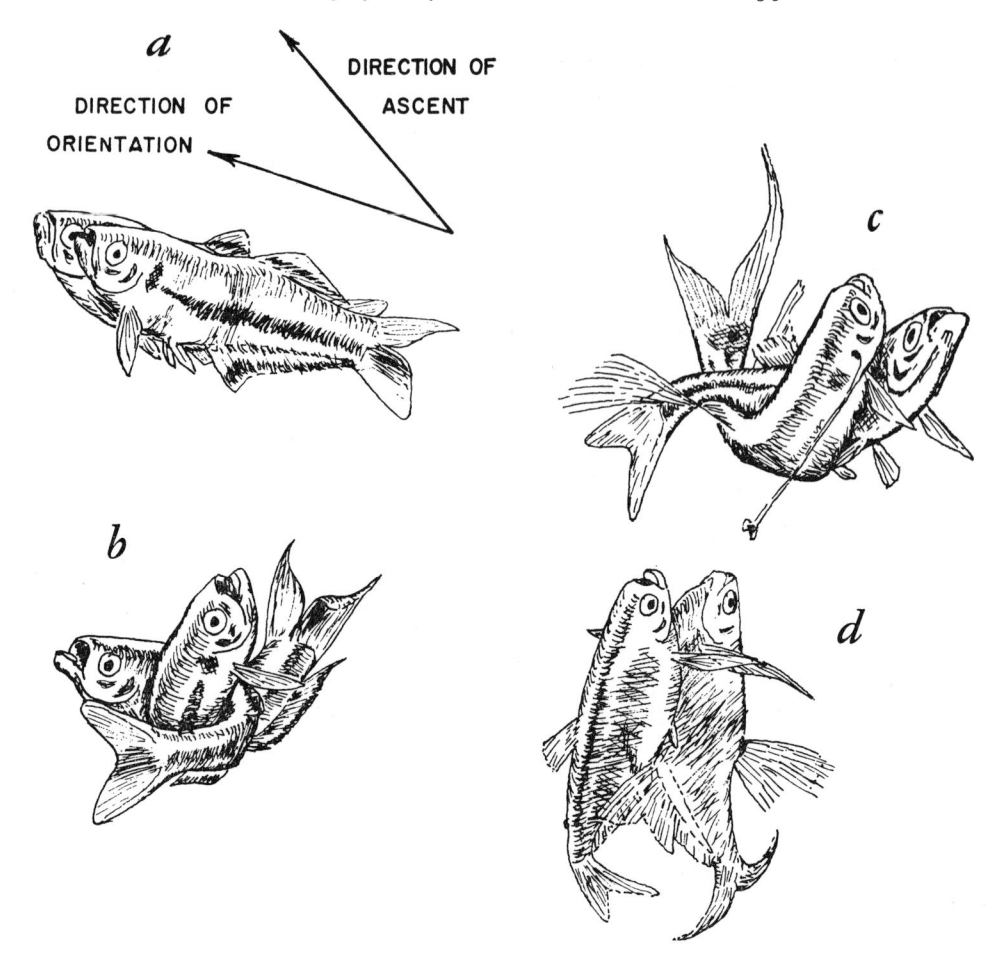

Fig. 11. Pairing, Rolling, and Impinging: *a*. Pairing in *C. microlepis*. The animal toward the observer is the male. (From photograph.) *b*. Rolling in *C. microlepis*. The female's head is toward the left. *c*. Rolling in *Corynopoma*. The right-hand head is that of the female. *d*. Impinging in *Ps. doriae*. The male is to the right. (*b*, *c*, and *d* from sketches. Explanation in text.)

right angles to it (fig. 7, *b*). The fins could be either erected or lowered. The Upright Posture was usually held for a second or longer.

In *Pseudocorynopoma* the Beat was not present. A male, when approached by another fish, suddenly would rotate his body 75–90° upward, and as quickly return to his resting orientation. Median fins were usually erected, and the mouth was often seen to be open.

Hovering (H).—This was seen in *Coelurichthys* and *Glandulocauda*. The male poised himself either above or below a slowly moving female, and following her movements precisely, remained there, often for several minutes at a time. As the female increased her speed, the male was evidently no longer able to mirror her movements and he dropped behind. This was used to distinguish Chasing from Hovering: if the male was above or below the female, he was considered to be

Hovering, if he was behind, or outside her when they swam in a tight circle, and the speed was sufficient it was called Chasing.

In all three species the male maintained himself in Hovering by constant fin and tail movements. It was observed that in all three, when Hovering Below, the male would orient normally, that is, the angles of pitch and roll were zero. However, when Hovering Above, male *Glandulocauda* and *C. microlepis* were seen to be pitched head downward at an angle of 15–25° (figs. 2, *b*, 8, *a*). In contrast, *C. tenuis* rarely Hovered directly above the female, but rather to one side, and was seen to be rolled in her direction, occasionally so much so that his sagittal plane was nearly horizontal (fig. 8, *b*).

Gulping (G).—During Hovering, the male *Glandulocauda* would nearly invariably dash to the surface and return with a bubble of air to Hover again above or below the female. Between Gulps the air would gradually or suddenly escape from behind his gill covers. The Gulps were regularly spaced, with a mean on different occasions of 6 to 20 seconds intervening between them. Regularly spaced Gulping was seen in no other characid studied.

Croaking (Cr).—During Hovering, but only after the first Gulp in a particular Hovering bout, the male *Glandulocauda* produced a continuous train of pulses of sound, each lasting from one to several milliseconds. The length of a train of pulses varied from less than one second to over 25 seconds, in which case it was momentarily broken by two or three Gulps. A further description will be given in a later section. Croaking sometimes continued briefly into Chasing if the female interrupted Hovering by the male. Once the male *Glandulocauda*, in a tank containing several *C. tenuis* (which were being monitored with the hydrophone), was seen to Nip Surface and produce a sound which did not appear to differ from Croaking. At no other time outside of courtship, and from no other species was a sound heard which resembled Croaking.

Zigzagging (Z).—Basically this was an upward movement followed by a downward movement. Each species possessed its own variant. In *Glandulocauda* the male, while Hovering, sometimes started toward the surface but returned to Hovering before he reached it.

In *Corynopoma* the male often Chased upward and then downward to position himself in front of the female; sometimes these Chases took the form of an inverted V. At other times he would describe a figure-of-eight in repositioning himself in front of the female. As all intergrades were found between these extreme variants and ordinary Chases, no attempt was made to record them separately.

Zigzagging in *C. tenuis* (fig. 9, *c*, 9, *d*) was more stereotyped than in *Glandulocauda* or *Corynopoma*. It occurred during Approaching, Chasing and Hovering, and consisted of one or more pivots, upward to an angle of 35–45° and downward to an angle of minus 20°. There was little tendency to swim upward and downward; Zigzagging in this species was often a change in orientation only, but it would often be repeated a number of times.

Zigzagging in *C. microlepis* displayed two relatively distinct levels of intensity. At the lower level, each cycle of up and down movement lasted about one second and consisted of three of four fused Beats directed 45–60° upward, followed by a quick turn and one or two downward Beats at 30–60° below the horizontal; the following turn into the upward movement was more gradual. The upward movement occupied about two-thirds of the cycle (fig. 9, *a*).

In the more rapid Zigzag, the cycles occurred at a rate of two per second; upsweep and downsweep each consisted of several fused Beats (fig. 9, *b*). The turn at the top of the movement was again more acute than that at the bottom. At each frequency the movement was graded in amplitude, and for the lower level varied from little more than successive changes in orientation, as in *C. tenuis,* to a movement in which the vertical distance from the bottom turn to the top turn was greater than two fish lengths. At the lower level, Zigzagging was usually performed in front of or alongside a female; at the higher level, the male circled about her.

In none of the species discussed thus far were the fins raised during Zigzagging. In *Pseudocorynopoma,* however, both the dorsal and anal fins were unfolded and the pectorals were held as in Wobbling, straight out from the body, and probably used as hydrofoils. The upward movement, usually at an angle of 60–90° to the horizontal, consisted of a variable number of lateral undulations, at a frequency of between three and ten per second. The downward movement was highly variable, and ranged from a long flat glide to a steep descent at an angle of 75° below the horizontal, and composed of several fused Beats. This account is based on *Ps. doriae; Ps. heterandria* did not appear to differ significantly (fig. 9, *e*).

Ovalling (O).—This movement was confined to *Pseudocorynopoma,* was only rarely observed in *Ps. heterandria,* but was common in *Ps. doriae.* The male swam around the female as indicated in figure 7, *c*. Usually he passed her in front below, and to the rear above her caudal peduncle, so that a line from the lower edge of her gill cover to the base of her dorsal fin would lie aproximately on the long axis of the ellipse described by the male. This pattern was upset, but recognizable, when obstacles such as plants, the aquarium wall, and other fish were in the way of the male. The dorsal and anal fins were partly to maximally unfolded, and his pectorals were actively used in locomotion as he passed above her, but not as he passed in front of her, at which time they appeared to be used as hydrofoils.

Jerking (J).—In *Coelurichthys* and occasionally in *Glandulocauda,* and in both especially during Hovering, males were seen to twitch or Jerk violently. The Jerk had the appearance of an extremely rapid Beat or sigmoid of small amplitude. This was also seen in *Pseudocorynopoma* occasionally during Approaching and Ovalling.

Vibrating (V).—This was confined to *Coelurichthys* males and was seen in Approaching and Hovering. The male gave a series of lateral undulations of small amplitude, at a frequency of eight (*C. microlepis*) to about thirteen per second (*C. tenuis*).

Dusting (D).—This movement was peculiar to *C. tenuis,* although indications of it were seen in *C. microlepis.* The male positioned himself in front and to one side of the female as in figure 8, *c*, with his body behind the dorsal fin bent sharply away from her. He then began to rapidly vibrate his caudal fin while straightening his body and thus bringing his caudal peduncle slowly back toward the midline. When it reached the midline he stopped vibrating it, and swung his peduncle rapidly back either to its original position or to the same position on the opposite side, usually the latter, and repeated the process. Each swing toward the midline lasted for about one to one and one-half seconds. The caudal fin was widely spread, but the dorsal and anal fins were folded. The effect was that of shaking a feather duster, hence the name. He rarely if ever seemed to actually brush the female's head with his caudal fin during this activity.

In *C. microlepis* males 14 to 18 weeks old, Dusting without the orientation in front of the female was often seen. This was never observed in fully adult males of the species, but on occasion fully adult males would Vibrate near the female.

The following four actions were confined to *Corynopoma*:

Extending paddle (E).—A male oriented parallel with, and somewhat in front of another fish, and Extended the opercular paddle closest to the recipient perpendicularly in front of the latter (fig. 10, *a*). Simultaneously the median fins were maximally unfolded, and often the performer Gaped. During Extension, which could last for only an instant or for as long as three seconds, the body remained relatively still.

Twitching paddle (T).—With fins raised, paddle extended and orientation as in Extending, the body was twitched at a frequency of about two per second. This appeared to produce a concomitant twitching of the paddle; the paddle movement appeared to be passive. Twitching lasted for up to five seconds and was again often accompanied by Gaping. Often during Twitching the male's body was bent, concave side away from the recipient.

Shaking paddle (S).—With body, fins, and paddle as in Extending, the median fins were folded over toward the recipient, and the male's body was set into a vibrating movement which in turn appeared to produce a passive flutter of frequency 6–10 per second in the paddle. Again, Gaping often accompanied this. In addition, the male often tilted his tail up so that it lay beneath the recipient. Shaking may last for up to two seconds (fig. 10, *b*).

Quivering (Q).—This began as a movement similar to Shaking, except that the paddle was not extended. The fin-folding about the recipient became more prominent, the tail was rotated so that the sword came to lie on the opposite side of the recipient, and the caudal peduncle was raised so that the recipient lay in the cleft between the caudal lobes (fig. 10, *c*). The male then began to accelerate in a forward direction as the quivering became of larger amplitude and developed into a fused series of Beats; with tail and enfolding median fins he appeared to be pushing the more-or-less passive recipient about the tank. Largely depending on the recipient's response, this movement lasted from one to six seconds.

Actions Performed Only or Predominantly by Females

Following (F).—This activity was really a form of Approaching, distinguished mainly by context. It occurred essentially as a *response* to an action by a male, and it has been thought best to give it a separate heading. It was seen in all species with the possible exception of *Glandulocauda*. In *Corynopoma* it occurred as a response to Extending, Twitching, and Shaking, and took the form of a slow turning toward and following of the paddle tip during these actions. It lasted for up to ten seconds (that is, through several male activities). During Twitching the male often turned slowly away from the recipient, and it was at this time that Following by a *Corynopoma* female was easiest to discern.

In *C. microlepis* the female often Followed a Zigzagging male; this took the form of a slight back-and-forth turning by her, similar to that of a tennis spectator. In *C. tenuis* it was seen as a response to a Dusting male; the female slowly moved toward him.

In both species of *Pseudocorynopoma* Following occurred as a response to Zigzagging; the female suddenly turned and moved toward the male during one of his upswings. It also occurred as a response to Wobbling by the male; the female turned toward him, Followed and then began to Wobble also. In *Glandulocauda,* Following, if it can be called that, took the form of a rising toward the surface below or above a Hovering, Croaking male.

Nipping paddle (N).—This occurred only in *Corynopoma*. While a male was engaged in Extending, Twitching, or Shaking, and usually while the female was Following the paddle, with a quick Beat she suddenly darted and bit at the tip of the paddle. A Nipping not preceded by Following often had the more violent character of Biting, as measured by subsequent male activity. Thus the distinction between Biting and Nipping rested on context.

ACTIONS PERFORMED JOINTLY BY MALE AND FEMALE

Pairing (P).—While a male *Corynopoma* was going through the complex that has been labelled Quivering, a female would occasionally cease to be passively pushed and would begin to vibrate actively. The male, which up to this time appeared to move her through the water only with effort, was suddenly seen to speed up. This combined quivering will be called Pairing.

Pairing was originally defined for *Corynopoma,* but similar acts were seen in *Coelurichthys* and *Glandulocauda,* except that the male's fins did not appear to enfold the female to as great an extent. It was seen only once in *Glandulocauda;* the fish were near near the surface and lined up as in a Lateral Display, fins folded, head to head. They then moved in circles at the surface of the water, no more than one-half cm. apart, and finally separated.

In *C. tenuis* and *C. microlepis,* the pair came together in midwater and rose to the surface along a line directed at 45 to 50° to the horizontal, but with their body axes directed upward at an angle of only about 15°, as in figure 11, *a*. In both *Glandulocauda* and *Coelurichthys* both members of the pair actively quivered. When *Coelurichthys* reached the surface, they circled in tandem, quivering, and very close together. In both species the bodies were closer together at the bottom than at the dorsal edge. On several occasions in both species of *Coelurichthys,* Pairing terminated by passing into Rolling (see below); at other times only one member of the pair, usually the female, leaped from the water. The great majority of Pairings ended by both fish simply swimming away from one another. In *C. tenuis,* as mentioned above, pairs of males in head-to-head Lateral Display occasionally passed into Pairing, rose to the surface, and circled in tandem. In all these cases Pairing was terminated by the animals swimming away from one another.

Impinging (I).—This action was found in both species of *Pseudocorynopoma*, but in no other Glandulocaudines. Occasionally while both male and female were Wobbling, they would simultaneously throw themselves into Upright Postures with their ventral surfaces in contact. Contact lasted at most for one-third of a second. No eggs appeared. Impinging is illustrated in figure 11, *d*.

Rolling (R).—This was seen in *C. microlepis*, *C. tenuis* and *C. riisei*. The Pairing fish suddenly intertwined and usually leaped as much as two cm. from the water. The movement was unanalyzable without motion-picture studies; however, by building up a picture from many observances, it was possible to construct figure 11, *c* for *Corynopoma*, and figure 11, *b* for *C. microlepis*. On one occasion the male *C. microlepis* returned to the water bent into an upside-down U, and hung in that position at the surface for several seconds, twitching spasmodically and with his mouth agape.

Defining an Action

In the study of behavior, problems arise in the recording of data which must inevitably affect their interpretation. In particular, the task of differentiating each behavioral act from its fellows, and the task of recording each act, pose two distinct but related problems.

The first is the practical problem of deciding when to consider an act as having begun and ended. When recording with a motion picture camera this may be decided at one's leisure; this is not possible when recording by eye, whether speaking into a tape recorder or manipulating the keyboard of a pen-writing mechanism. In these cases one should be equipped with rigid criteria enabling one to determine reflexly what constitute behavioral acts; these criteria should be communicable.

The second problem is more theoretical and involves the fact that the onset of many behavioral acts is gradual rather than sudden. In these cases it is difficult to place the beginning of an act in time, even with motion picture recording. If the act also ends gradually, it may even prove impossible to count acts of the same kind when they follow one another without pause. Considerations of this sort led to the development of the following criteria and methods.

In recording the behavior of *Corynopoma*, a tape recorder was used until August 1961. Subsequently, an Esterline-Angus 20-channel operations recorder with a keyboard was used to record all timed observations. Later observations were also monitored with a hydrophone; Croaking was thus recorded both on one Esterline-Angus channel and on magnetic tape.

In the earlier observations, a code letter was spoken into the recorder at the beginning of each distinguishable activity; these letters were subsequently transcribed to mechanics' bond paper on which each division stood for five seconds of time. Intervals were measured from the middle of one letter to the middle of the next and were thus regarded as intervals separating instantaneous events. With the Esterline-Angus operations recorder, each channel represented one kind of action; as the number of kinds of event rose beyond twenty, pairs of keys representing incompatible actions, for example Biting and Nipping Surface, were used together to represent a third incompatible event, such as Yawning. The chart drive was run at three inches per minute and intervals were recorded to the nearest second. Lengthy actions were recorded in full, and were not regarded as instantaneous. The point of jotting down the code letter in the earlier records thus corresponds to the *beginning* of the recorded event on the operations recorder.

It remains to give the criteria used in defining the beginnings and ends of these events. Observe that in order for two or more events of the same class to be considered as separate from one another they must have clearly definable beginnings or endings.

Such activities as Approaching, Chasing, Following, and the like were treated as having indefinable beginnings, unless immediately preceded by definable endings of previous events. They were recorded as soon as they were *evident*. They were regarded as ending when the performer turned away or when a definable beginning of another activity supervened, or, in the case of male courtship Chasing in *Corynopoma*, when a stable orientation in front of the female was achieved. The criteria for Hovering and Chasing in *Coelurichthys* and *Glandulocauda* were given in the preceding section.

Such actions as Extending, Dusting, Lateral Display, Wobbling, Zigzagging, and Croaking had easy-to-define beginnings and ends and need not be dwelt upon. Zigzagging in *Coelurichthys* was recorded throughout the cycle, with a break at the bottom of each cycle. In *Pseudocorynopoma* it was recorded only during the upward movement, with a pulse at each undulation. This procedure was followed also in Wobbling.

A record of Twitch or Shake was begun at the first sign of it, and ended if more than about a half-second elapsed with no Twitching or Shaking.

Upright Posture in *Pseudocorynopoma*, Biting, Nipping, Nipping Surface, Gulping, Impinging, and Tail-Beating were regarded as instantaneous events. In *Glandulocauda*, Gulping was not regarded as interrupting a bout of Hovering.

Those acts which have in themselves no definable beginning or end (such as Following, Approaching, most Chasing) are also the longest; it may be that although the observer cannot subdivide them, they are quantified for the performer.

A special case was the driving of the female in a circle exhibited by *C. tenuis* males, considered in other respects to the Chasing. In this case each circuit was recorded as a separate movement.

PATTERNS OF SOCIAL BEHAVIOR

Some Inferences and Definitions

Courtship behavior.—With organisms in which fertilization is external it is usually possible to define and measure the "success" of "courtship" activities by reference to the spawning act itself; with animals in which internal fertilization occurs this is usually not so simple. For example, Beach (1954) has pointed out that in many mammals, "copulation," "intromission," and "ejaculation" are anything but synonymous. Similarly, only by combining observations with genital smear techniques were Clark and Aronson (1951) able to demonstrate that in the guppy (*Lebistes*) only copulations lasting at least 0.8 seconds resulted in transferral of sperm. They added that female cooperation was necessary, and that most previous students of *Lebistes* had probably never seen or recognized a true copulation.

Breider (1948) has established that in *Corynopoma*, "courtship" results in the deposition of spermatophores in the female's genital tract, but unfortunately he did not state how much of the courtship pattern was required for this. Again, Kutaygil (1958) established that presence of the paddles in *Corynopoma* was not absolutely essential for fertilization of the female to occur, but that without an intact anal fin, the male was unable to inseminate the female. He reasoned that the anal fin hooks were the requisite feature. Again, he gave no behavioral evidence. In the present study histological investigation proved impractical for lack of sufficient material, and tank space in which to isolate virgin females was at a premium.

Therefore, in the absence of conclusive evidence that a particular act results in sperm deposition, the following argument and inferences are presented *Corynopoma* will be considered first.

1. Females isolated from before sexual maturity, when placed with males but removed before the occurrence of Quivering, Pairing and Rolling, did not spawn. This implies that other activities seen in these encounters do not result directly in insemination.

2. Females, isolated as before, and left with males until after Quivering, Pairing and Rolling, did occasionally spawn. It is inferred that one of these must result in sperm deposition.

3. A view often expressed in the hobbyist literature is that the male's opercular extension is used to transfer spermatophores to the female, either via her urogenital sinus (Hoedeman and de Jong, 1949) or via her mouth (Fraser-Brunner, 1948). Both interpretations would require the paddle to be bent into a semicircle to receive a spermatophore from the male's vent. A histological study of the opercular extension revealed no muscular attachments which could bring this about. Neither interpretation can explain Kutaygil's finding that fertilization could occur in the absence of both paddles.

4. As there are no other evident external genitalia aside from the female's ovipositor, which only appears at spawning, it is inferred that spermatophores must be passed directly from vent to vent. This presumably would have to be by direct contact or by very close proximity, and this does not appear to be possible during Quivering and Pairing. It would therefore appear that insemination must occur during or after Rolling. As the fish separate after Rolling, spermatophore deposition must occur during Rolling.

Rolling occurs so rapidly and so infrequently that it has not been possible to obtain the necessary photographic evidence to determine whether vent contact actually occurs.

In the early series, fourteen good records, of which six were tape recorded, were made of events leading up to Rolling. With one exception, in these records Rolling followed a fairly rigid set of prior actions, namely, Chase, Extend, Twitch, Follow, Nip, Quiver, Pair, in that order (the number of events of each kind could vary, but the order was very nearly the same in all). Subsequently, the same order appeared in many records. Furthermore, these eight acts were either confined to or far more frequent in male-female encounters; all other social actions except Shaking were as common or more common in other contexts.

Therefore in *Corynopoma* these eight classes of events are classified as Courtship Activities: male Chasing, Extending, Twitching, Quivering, female Nipping and

Following, and male and female Pairing and Rolling. In addition, Shaking, which was absent during most of the series preceding Rolling, but was far more evident during male-female encounters than in other contexts, is included also as Courtship Activity.

Of these, Extending, Twitching, and Shaking are grouped together as male Displays, and Following and Nipping as female Responses. In the following sections concerning *Corynopoma*, male Displays and female Responses have no other meaning. The word "response" ordinarily has a cause-and-effect connotation which is unfortunate, but no other term seems to convey the intended connotation that these acts were seen only in the presence of male Displays. The use of the term "response" is justified elsewhere by the demonstration that male Displays have a differential effect upon the occurrence of Following and Nipping (Nelson, in press). Altmann (1962, page 381) discussed the propriety of including a specified antecedent in the definition of an act.

Actions which appeared to be quite similar to Pairing and are here so called have been seen in both species of *Coelurichthys* and in *Glandulocauda*. An act similar to Rolling in *Corynopoma* was seen in both species of *Coelurichthys*. In all three, Pairing was preceded by extensive Hovering and Chasing. In *Glandulocauda* it was preceded by occasional Zigzags and Gulping. Probably Croaking also occurred, as it is nearly invariably associated with Gulping and Hovering. In *Coelurichthys* Pairing and Rolling were preceded by extensive Zigzagging. In *C. tenuis* Pairing and Rolling usually followed Dusting most closely in time. Vibrating and Jerking were seen only in association with the above events. Female Following was defined by its association with Zigzagging (*C. microlepis*), Dusting (*C. tenuis*), and Hovering (*G. inequalis*). Because they occur during long periods of no overt aggression (Biting), and by analogy with *Corynopoma*, all the above activities are defined as Courtship Activities. The term Display is used in conjunction with these activities, and the term Response is confined to female Following.

On one occasion a male *C. microlepis* returned to the water after Rolling still bent double, having released a small amorphous mass. This was secured with a pipette and proved to contain non-motile sperm in a curd-like matrix.

For *Pseudocorynopoma* the reports in the literature are conflicting. Sterba (1959) placed them in the category of *Corynopoma*, but A. V. Schultz (1962) described the *Ps. doriae* pair as scattering eggs. During the present study neither species was observed to spawn. However, Ovalling and Zigzagging in these species bear resemblance to similar acts in the others, and Wobbling appears to be a continuous Zigzag with a horizontal direction of movement. Impinging is unique, but is temporally related to mutual Wobbling as Rolling is to Pairing in the other species. In addition, Approaching by the male was predominantly associated with Ovalling and Wobbling, and Following by the female was previously defined by its association with Zigzagging and Wobbling. By analogy, therefore, the above temporally associated activities in both species of *Pseudocorynopoma* are defined as Courtship Activities. Female Following and Wobbling are termed Responses, and for these species this term has no other connotation.

Aggressive Behavior.—The criteria used to define Aggressive Activities were (a) correlation in time with Biting, and (b) the presence of large blocks of the behavior in question without, or with only sporadic occurrence of Courtship Activities.

In *Corynopoma*, Chasing and Biting were usually the only aggressive manifestations. Occasionally in male-male encounters the Upright Posture was seen in situations which indicated that it was probably aggressive in nature. No analysis could be made of the rare Lateral Display in this species, although it was evidently morphologically related to the Upright Posture.

In *Pseudocorynopoma* of both species the Lateral Display and Upright Posture were associated in time with Biting, and will be considered Aggressive Activities. Chasing was only rarely seen outside of an aggressive context.

In *Coelurichthys* and *Glandulocauda*, Aggressive Activities include Biting, Lateral Displays, Chasing (when followed by overtaking and, usually, contact), Tail-Beating, and Head-on Orientation, all as judged by their temporal association.

Behavioral Units.—Actions have been defined in a previous section. A *bout*, with the exceptions to be mentioned, is here defined as an uninterrupted grouping in time of acts of the same class, e.g., Chasing, or Dusting. It is understood that an act by a participant other than the performer does not of itself interrupt the latter's bout. The exceptions are all in *Glandulocauda,* and all involve the complex of Hovering, Gulping, Zigzagging and Croaking, usually performed together. A bout of Hovering or Croaking was not considered to be interrupted by Zigzagging or Gulping; however, any other break, even a few silent or non-Hovering seconds, was considered to break the bout. The bout measure was not used with Gulping and Zigzagging in this species.

A *sequence* is defined as a grouping in time of events of related classes, e.g., a Courtship Sequence. A *series* is any series of behavioral events in time.

Social Interactions in Non-Courting Animals

In this section will be discussed the behavior of homosexual and heterosexual groupings in which long courtship sequences do not supervene.

Behavior of the Young

The behavior of immature fish was similar in the three species studied, *Corynopoma, C. microlepis,* and *G. inequalis*. In all three the eggs hatch in 20 to 36 hours and the young remain hanging by means of a frontal adhesive organ for several days. Before the yolk sac is completely absorbed they are free-swimming, and soon move freely through all levels.

The development of social interactions will be described first for *Corynopoma.* Aggregating did not appear spontaneously, nor could it be induced, in young under three weeks of age. After this time the young could be made to aggregate (for example, by jarring the tank) but the distance between fish relative to their length was greater than in aggregated adults until the young were nine weeks of age. Aggregating in the young *Corynopoma* was centered in and about the largest clump of plants in the tank; if two clumps were present, two aggregations might form. After 12 weeks, this association with plants was no longer to be observed. In a 30-gallon tank, 10 inches deep, juveniles occupied the upper strata where adult females rarely ventured; no such stratification was seen in juveniles alone.

When newly hatched *Artemia* were fed to juveniles the latter aggregated in the densest part of the cloud of shrimp, and at a little over two weeks of age the young were observed to nip at one another while feeding. After five weeks of age, nipping developed the character of adult Chasing and Biting, and also was seen (at a lower frequency) in non-feeding situations. Dropping was never seen in young of less than 12 weeks of age. The Upright Posture was observed very occasionally in young more than seven weeks old, in which it always occurred upon the Approach of another fish. In *Corynopoma,* sexual differentiation was noticeable at the age of 17 weeks.

In *C. microlepis* the same changes were observed but with a somewhat more accelerated time course, so that Chasing and Biting occurred at the age of a month, and signs of sexual differentiation were seen at 10 weeks. Lateral Displays appeared concurrently with sexual differentiation. Aggregation of nearly adult form could be produced at the age of three weeks.

In *Glandulocauda* the rate of development was even more accelerated, so that signs of sexual differentiation and Lateral Displays were seen at an age of seven weeks. At this time aggregation was still difficult to produce; however, this may have been a function of the small size of the group (six young).

THE STABLE GROUPING

In groups of adult females *Corynopoma* and in the heterosexual group of *Ps. heterandria* which had been together for some time, behavior changed little from day to day, and a sort of equilibrium appeared to have been established. This will be referred to as the stable grouping; in male or heterosexual groupings of *Corynopoma* and in all groupings of adult *Coelurichthys* and *Glandulocauda* the stable grouping is an abstraction.

The stable grouping was characterized by more rapid movement, more frequent turning and more frequent Nipping Surface than in isolated individuals. Sporadic aggregation was seen, at intervals of several minutes to several hours; *C. tenuis* was aggregated almost continuously. Aggregation could be induced by the approach of an observer or by jarring the table upon which the tank rested, but also occurred in the absence of evident change in external stimulation.

Except in *C. tenuis,* few Chases and Bites were observed in aggregations. Simultaneous with the beginning of dispersal of the group in *Corynopoma,* one or more females were seen to begin to Chase and Bite one or more of the others; the general diaspora appeared to result largely from this. In this species Chases and Bites recurred throughout the periods of dispersal. *Ps. heterandria* was similar in these respects to *Corynopoma.*

In *C. microlepis, C. tenuis* (on the rare occasions on which the aggregation was seen to break up), and *G. inequalis,* a rise in frequency of Lateral Displays occurred concurrently with the increase in frequency of Chases and Bites at the beginning of dispersal. Both sexes took part in these actions.

When aggregated, the total group moved from place to place less than did individual non-aggregating fish. Further, the area preferred by the aggregation remained constant for weeks at a time, and appeared uncorrelated with the degree of plant cover. In contrast to the behavior of juveniles, Chasing, Biting and Lateral

Displays were less prevalent when food was plentiful than when it was scarce. Dropping was not seen in *Pseudocorynopoma, C. tenuis,* or female *Corynopoma* groupings.

TRANSIENT BEHAVIOR

When a fish was placed in a tank with a resident individual or group, a certain amount of time was required before the conditions of the stable grouping supervened. As the encounter technique was used in the analysis of courtship behavior in several of the species studied (Nelson, in press), it was necessary to understand the changes which occurred in non-courtship situations upon introduction of a fish.

If an animal were placed with another previously isolated individual, the usual first response of both was to Drop. If the introduction were made gently enough, only the newcomer Dropped and the resident usually Approached. Little further occurred until the newcomer began moving about the tank; further behavior depended upon species and sex of both participants.

Most typically, the two would Approach and circle one another; this had the appearance of aggregation rather than of "exploration." With female *Corynopoma,* Chasing and Biting might follow; the identity of the Chaser could not be predicted from previous behavior. If Chasing and Biting did occur, they were more severe and frequent than in the stable grouping, and the Bitten female would frequently Drop. On the three occasions that a female *Corynopoma* was placed with a group of two resident females in which stability had been established, the introduced female proceeded to Chase and Bite the others; this was neither frequent nor too severe, to judge by the mild reactions of the residents. If a female were placed with a larger stable group her initial Dropping was of shorter duration and she soon joined the aggregation if it had formed. The stable grouping was not upset by her introduction and the behavior of the newcomer was soon indistinguishable from that of the residents.

Often in *Pseudocorynopoma* and *Corynopoma,* male-female encounters took this same course, if there were no courtship. In *Corynopoma* male-female encounters, if Chasing and Biting occurred, the female was nearly always the aggressor and the male often Dropped. In *Pseudocorynopoma,* the male was most often the aggressor, and the Chased and Bitten female rarely Dropped.

Addition or removal of an animal from a small group of *C. microlepis* caused radical changes in social interactions, which will be discussed in detail later. In *C. tenuis* the results were similar to those with *Corynopoma,* except that complex series of Lateral Displays occurred, to be described in a later section. In *Glandulocauda,* if Lateral Displays did not occur, non-courtship encounters always ended with the larger animal Chasing and Biting the smaller, which Dropped.

MALE-MALE ENCOUNTERS IN CORYNOPOMA

It is impossible to point to a "typical" record of an encounter between two males of this species. Most of the components seen in male-female encounters were observed, but they were less highly organized and less predictable. To demonstrate the sort of changes which may occur over a period of several hours, the frequency-time graphs of figure 12 were prepared for Upright Postures, Approaches, Chases, Bites, male Displays, Quivers, and other measures, from records of alternate half-hours of a 7½-hour encounter between two males. For each activity, frequency per minute is given above the horizontal line for male 1, and below it for male 2. Figure 12 will serve to illustrate the following generalizations.

1. Social interactions were infrequent in male-male encounters compared with other species studied. Thus, in figure 12 there is an average of only three or four social activities per minute per fish.

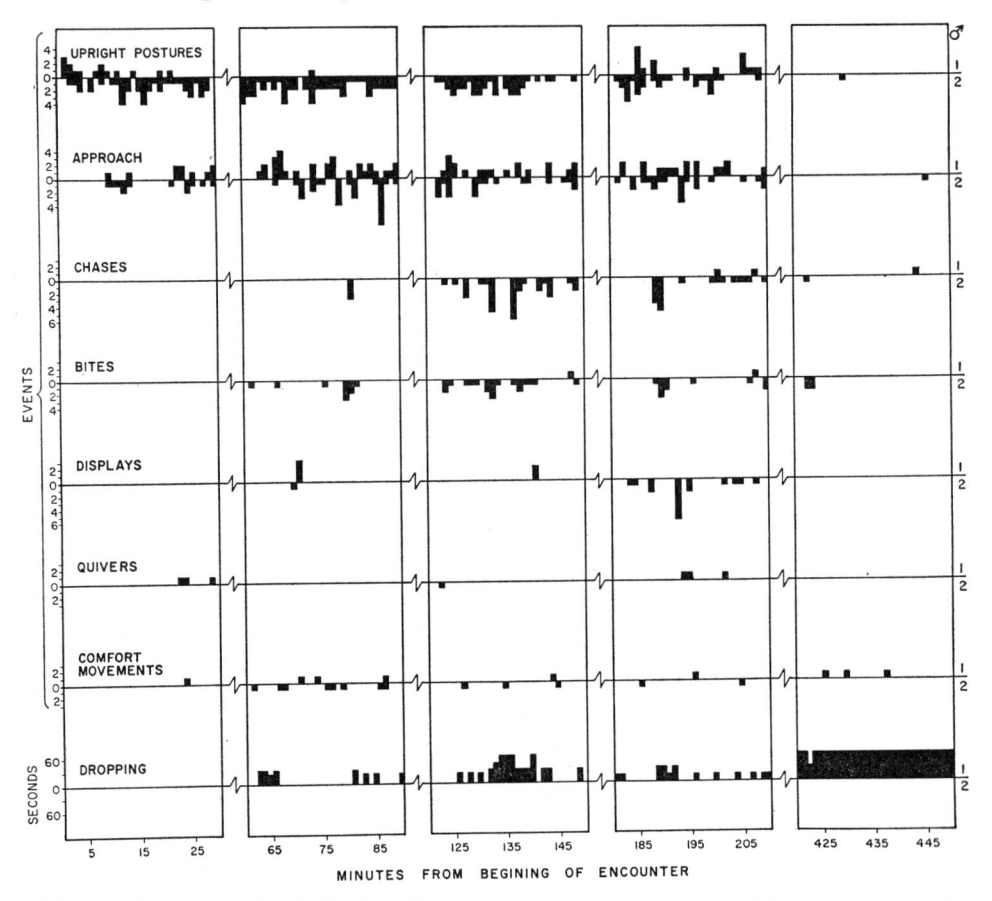

Fig. 12. Frequency of activity in a *Corynopoma* male-male encounter. The ordinates are frequencies of acts per minute and for Dropping, the number of seconds of each minute which were spent Dropping. Five half-hours, the first, third, fifth, seventh and fifteenth, are shown. The records above each line are for Male 1; those below, for Male 2. (Explanation in text.)

2. When one animal was doing something, the other was usually doing either something else or nothing. This is perhaps most clearly shown in the Upright Posture record of half-hour, and in the Approach record of half-hour 3. This is in contrast to encounters in *Coelurichthys* and male *Pseudocorynopoma*, in which both animals tended to be engaged in the same activity at the same time.

3. As Upright Postures became less frequent, first Chasing and Biting and then Dropping became more frequent. Exceptions to this rule occurred, however, in

which Upright Postures and Chasing peaked together, then fell as one animal came more and more to Drop.

4. Displays and Quivering were always related inversely to Upright Postures by the same animal. The record of figure 12 shows them also to be inversely related to Chases and Bites, but this was not invariable.

5. Bites always peaked with Chases, for obvious reasons, but not vice-versa.

6. Quite often, one male would Posture more in the early part of the observations than later. In this encounter, Male 1 Upright Postured more at the very beginning of the period, but the balance soon shifted to Male 2.

7. This record offered an insight into the relation between Upright Posturing, Biting and Chasing which hitherto had been puzzling. In half-hour 7, Male 1, who had been repeatedly Chased and Bitten during the previous hour and had been seen to Drop repeatedly, was left alone by Male 2 for nearly 10 minutes. Soon after the beginning of the seventh half-hour, Male 1 indulged in a resurgence of Upright Postures, whether or not Male 2 Approached. After a little of this, Male 2 Chased and Bit repeatedly at Male 1, who Dropped to the bottom. Many later incidents of this sort led to the conclusion that the upsurge of Posturing by the one was directly responsible for the increased Chasing and Biting on the part of the other, although the temporal correlation of these Aggressive Activities was not as close as in *Pseudocorynopoma*.

8. After a period of from two to ten hours, one male would spend nearly all his time at or near the bottom, or swimming up and down the side. He would be sporadically Chased and Bitten, and occasionally Postured at, by the other. This is illustrated by the last half-hour in figure 12.

9. After several days, the Chaser and Biter would still be the same individual; Chases and Bites would occur in short bursts from 20 minutes to two hours apart. Dropping by the Chased individual would then be less frequent and less prolonged than early in the encounter. Among groups of males in 35-gallon tanks, behavioral stability of a sort supervened, punctuated by Displays and Quivering. Often one male spent much time Dropping as a result of frequent Chasing and Biting by another individual in the group.

10. In every case, a male destined to Chase and Bite would be more likely to Posture than would one destined to Move Away rapidly and to Drop. In no case were fish observed to reverse roles once these had become established.

In 16 male-male encounters the effects of an animal's previous experience was noted. On several occasions a fish which had been reduced to continuous Dropping was placed with a male who had had recent opportunity to court females. The Dropper was courted at first, and on two occasions was soon Chased and Bitten. On the third of these the Dropper Postured upon the Approach of the Courter, and was soon Chasing and Biting the latter. These and other encounters are summarized in table 3.

Several records were made of three males placed together. In these, one male came to be the most frequent Biter; surprisingly, he had been a Dropper in encounters between two males. Attempts to change his position by removing him, allowing another male to become Biter, and then replacing the previous Biter, were without success. This trio had the appearance of a straight-line dominance hierarchy, with one most frequent Dropper, an intermediate Biter that Chased and Bit the Dropper, and a dominant Biter that paid little attention to the Dropper.

MALE–MALE ENCOUNTERS IN PSEUDOCORYNOPOMA

When two male *Ps. doriae* were placed in a 15-gallon tank, a high degree of aggressive behavior, Lateral Displays interspersed with Chasing and Biting, invariably resulted. The invariable conclusion was that one animal was Bitten so badly that large patches of scales were removed from his flanks; to save him, he had to be removed. The typical time course was an initial rise to a peak frequency of Lateral Displays of four to six per individual per minute, the gradual interpolation of

TABLE 3
Effect of Previous Experience Upon a Male's Role in a Male-Male Encounter
(Numbers in parentheses refer to the number of observations.)

An animal whose prior experience was as:	Becomes, when placed with an animal whose prior experience was as:		
	Courter	Biter	Dropper
Courter	Biter (3) or Dropper (3)	Dropper (3)	Biter (2) or Dropper (3)
Biter	Biter (3)	Biter (2) or Dropper (2)	Biter (3)
Dropper	Biter (1) or Dropper (2)	Dropper (3)	Biter (2) or Dropper (2)

Bites, and finally (after 30 to 45 minutes), the dropping out of Lateral Displays, Approaches and Biting by one individual and their increase to a total frequency of seven to eight per minute by the other. Few of these encounters were studied because of the limited amount of material available. Upright Postures as defined for this species were only occasionally seen, but there was a tendency for both animals to end a Lateral Display with their heads tilted upward. This was especially frequent in the animal destined to become the Bitten, who also showed fewer Approaches during the early part of the encounter. If there was an appreciable size difference, the larger animal became the Biter, but as only four mature males were available for these encounters, this is not conclusive.

On one occasion this same behavior was seen in an encounter between a male and a female; the latter emerged as the Biter. In all other male-female encounters, several times with the same male as in the just-mentioned occasion, little aggressive behavior was seen, and the performer was nearly always the male when it did occur.

When these animals were being kept together in a 30-gallon holding tank, one animal would usually be hidden in the *Fontinalis* cover provided; aggressive behavior between the other males was sporadic, and they were seen to remain spaced out.

Aggressive behavior in *Ps. heterandria* consisted of the same elements. It was observed only in a 50-gallon tank, where the fish generally aggregated peaceably, moving more or less as a group. Occasionally one animal would Bite at another, who would Lateral Display in front of the Biter.

In both species, interspecific Chasing, Lateral Displays and Biting could be elicited by members of both sexes of the other *Pseudocorynopoma* species, large *Corynopoma* and *C. microlepis* of both sexes, and two-inch *Hemibrycon* and *Knodus* (*Ps. heterandria* only tested). These interspecific encounters were not further explored. However, *Ps. doriae* males Chased, Bit and Lateral Displayed more often at male than at similarly sized female *Corynopoma* and *C. microlepis*. These did not appear to elicit the behavior, and responded by Moving Away.

AGGRESSIVE BEHAVIOR IN GLANDULOCAUDA

As only one male was available until late in the study, examination of aggressive behavior in this species has been confined to male-female encounters and to the behavior of larger groups.

Biting in this species was strictly size-dependent, as far as could be determined. The large male Chased and Bit, but rarely Laterally Displayed to, all others of its species with impunity, and only occasionally was Lateral Displayed to or Bitten by the others. The latter invariably suffered as a result of their transgressions. The largest female was in turn rarely bothered by other animals. In a tank containing two of her recently differentiated sons, she patrolled one area. Neither of them was able to court her. As he grew older (at 10 weeks) one of the young males also took up a sort of quasi-territory, and was occasionally seen to Chase others of his and other species out of it. In both cases the quasi-territory was an area defined by but free of plants, and containing about 60 cubic inches of water. When the old female and her son were driven out of their respective areas by the observer, they sometimes encountered one another and Lateral Displayed. On these occasions the female Chased the male about the tank, and then returned to her area where she swam rapidly back and forth for a time. This quasi-territoriality was seen also in *Coelurichthys*, especially *C. microlepis*.

In the bare tank used for experiments on respiration in *Glandulocauda* (Nelson, 1963), the male spent much of his time driving the female about the tank, Chasing and Biting at her. She would often respond to this by Dropping, but equally often by frantic swimming up and down the side of the tank, or by rising to the surface in a corner and remaining there. Bouts of remaining at the surface were generally shorter than bouts of Dropping, and both were usually broken by the Approach of the male and renewed swimming up and down the side. As mentioned, Lateral Displays by this male were rare; he often began an encounter with one of these smaller animals by Chasing and Biting. However, on several occasions he responded to the presence of a large female *Corynopoma* by prolonged bouts of Lateral Displays, in front of and alongside the female, who ignored him. The male *Corynopoma* elicited occasional Chases and Lateral Displays.

AGGRESSIVE BEHAVIOR IN COELURICHTHYS

Comparison of aggressive encounters.—In both species, nearly continuous series of aggressive actions were recorded in both homo- and heterosexual encounters. In both species these were nearly invariably present in encounters between males. In females of *C. microlepis*, these long series became more evident near spawning time, but were quite common as well at other times. In females of *C. tenuis* they were much more rare. Females of both species reacted to the presence of males with Lateral Displays, and the males usually responded in kind. If in these encounters the male turned out to be Biter, his behavior most often passed over into courtship. For a time the female would occasionally respond to courtship with Lateral Displays, fins folded, and these would often evoke renewed Chasing and Biting. If the male became the Bitten, the female Chased him mercilessly about the tank, and courtship never ensued.

Although both sexes of both species were observed to perform all four orientations in Lateral Display, with fins both raised and folded, differences in relative frequency were observed both between sexes and between species. In both sexes of *C. microlepis*, Circling was the most frequent orientation. In males the Broadside orientation was next most frequent, and this was followed by the Double Broadside, rare in female-female encounters, and the Head-to-Head orientation, rare among males and nearly absent among females.

In *C. tenuis* males on the other hand, the Head-to-Head orientation was the commonest, followed by single Broadsides, Double Broadsides and rare Circlings. In female-female encounters, single Broadsides were most frequent, but the Head-to-Head orientation was slightly more common among females in male-female encounters. Double Broadsides and Circling were rare among females. The Head-on Orientation was only rarely seen, usually preparatory to Double Broadsides, in *C. tenuis*, but was found for minutes on end among *C. microlepis*. Tail-Beating on the other hand was much more common among *C. tenuis*. This comparison is summarized in table 4. In females of both species, the fins were folded more often during Lateral Displays than in the males.

At the peak of aggressive encounters, both sexes in both species had generally darker markings. These lightened in fish which Dropped and in males as they began to court. As mentioned earlier, in *C. tenuis* the Head-to-Head orientation often passed into an act indistinguishable from Pairing. This occurred in both homo- and heterosexual groupings.

In a 30-gallon tank containing a large group of 16 week old sub-adults, individuals of apparently both sexes were seen to leave the upper waters occupied by the group and take up areas in clear spaces near the bottom. When other fish Approached them they were Chased and Bitten. If the tank were jarred these individuals joined the aggregation, but returned to their areas before the rest of the aggregation had broken up. They usually did this with Biting. One young male was seen to occupy the same small area over a period of several weeks. In other tanks, each containing several adults, similar behavior was often seen and was combined with dominance relations in a way to be described in the next section. A further elucidation of this phenomenon of quasi-territoriality will be given in a later paper.

Experimental alteration of social relationships in C. microlepis.—In this species, remarkable changes in behavior could be produced artificially by addition or removal of an animal from a group. For these experiments four males, 1, 2, 3 and 4, and four females, A, B, C, and D, were used in various combinations in 15-gallon tanks. The size relations among these animals were as follows:

$$4 > 2 > A > 1 > C > 3 > B > D$$

The differences between 2, A, 1, C and 3 were slight. The sorts of changes which could be induced are described in the examples in the following paragraphs.

1. Males 2 and 3, and Females B, C, and D were in tank 4. Male 2 was courting C; 3 was courting B; courtship was nearly continuous. D swam about in the left side of the tank. No aggression was seen save an occasional Bite at D by one of the other females. Male 2 was removed. Immediately 3, B and C began Lateral Displaying; usually 3 or C initiated the displays. Lateral Displays involving all three near the center of the tank were common. Biting, usually by C, began to occur. Each fish soon took up a quasi-territory and retained it, with occasional departures, for an hour. The following day the areas occupied were the same. Male 2 was reintroduced and

TABLE 4

Relative Frequency of Aggressive Actions in *Coelurichthys*
($++++$ = common; $+$ = rare)

	Lateral Displays: Circling	Broadside	Double Broadside	Head-to-Head	Head On	Tail Beat
C. tenuis:						
males.......	+	+++	++	++++	(+)	++
females.....	(+)	+++	++	+++	(+)	++
C. microlepis:						
males.......	++++	++	++	(+)	++	(+)
females.....	+++	++	(+)	(+)	++	+

immediately Approached C, who Broadsided. Male 2 Bit at her, Chased, and Hovered. Five minutes after 2's re-entry 2 and 3 were actively Zigzagging by C and B respectively, and the situation was not detectably different from that prior to 2's removal on the previous day.

2. Female A and Male 1 were in tank 3. Male 1 was sporadically courting to A, who responded with Broadsides, but did not Bite at him. Male 4 was introduced; both 4 and A Broadsided and Bit at 1, who Dropped. Male 4 and A alternately Circled, Broadsided and were stationary in the Head-on orientation for 20 minutes, but did not take up areas. Male 4 was removed, and 2, who had been actively courting in tank 4, was introduced. After two Circles, and a Bite by Male 2, Male 2 and Female A each took up an area and 1 remained Dropped in a corner. Upon removal of Male 2, A and 1 began again to Circle, and 1 was soon courting a Broadsiding female, as before.

3. In tank 3, several days before the events described in the previous paragraph, 2, 1 and A were occupying areas as before. Male 1 was removed; A and 2, who had not been Lateral Displaying at all, now began to Circle in the middle of the tank. 2 began to Bite, and soon A was Moving Away rapidly and Dropping. After a time 2 began to court, but then he began to Chase her, to which she responded by Moving Away rapidly and by Dropping. Male 2 was removed, and Male 4 introduced. At first he Dropped and A swam rapidly back and forth over the middle of the tank. Then 4 rose and Chased A out of the middle of the tank which he began to patrol himself, although he had not previously been in tank 3. Soon, without having Lateral Displayed or Bitten, he began to court most intensively. This ended in Rolling.

These examples should suffice to demonstrate the following facts. First, dominance, as measured by immunity from Biting, the ability to drive another individual from an area, and the degree to which Approach elicited Dropping, was found to be closely correlated with size. Secondly, the quick establishment of dominance by a male over a female appeared to be necessary (and in *C. tenuis* and *G. inequalis* too) for later successful courtship. Third, although dominance could be predicted from the relative size of the fish, the *form* of the social interactions, whether quasi-territories would be established, etc., could not be predicted with accuracy. Fourth, even so, once a particular group had been observed, their behavior could be predicted with great accuracy. Thus, Males 2 and 1 and Female A together in tank 3 *always* occupied the same areas, but the situation of Female A and Male 4 alone in the same tank always resulted in courtship. Fifth, in contrast to *Corynopoma*, the immediately previous experience of an animal had little effect upon his performance in a new group. Thus, with Male 1 and Female A, Male 2 patrolled the middle of the tank regardless of whether he had previously been courting or Lateral Displaying and Biting. Sixth, there were strong signs of accurate individual recognition in this behavior. Once a particular group had become established, reëstablishment of that relationship in that group in a new encounter was much more rapid, sometimes nearly immediate. The duration of this memory was not explored. Strong evidence of individual recognition was also seen in courtship in this species, and will be presented in a later section. Finally, the establishment of relationships in a new grouping were with few exceptions, in both this species and *C. tenuis*, begun with an extensive period of Lateral Display. The exceptions mostly concerned Male 4, considerably larger than any of the others, who often began immediately to court and was generally not responded to with Lateral Displays.

COURTSHIP IN CORYNOPOMA

Description of an encounter.—Figure 13 is a timed record taken with the tape recorder of the behavior which ensued when a male was introduced into the aquarium of a resident female. It is atypical in that social activity is more closely aggregated in time than is usual, and in that all courtship actions appear.

The male was introduced at 2:30 P.M. and after a brief Drop, Chased (C) the female[5] and soon Extended (E) his paddle in front of her; he then Twitched (T) it, whereupon she Bit hard at it (B). At the beginning of the second minute, he again Approached and Extended, then Twitched his paddle; she Followed (F), Nipped (N) at it. After sporadic Bites by the female and Extensions and Chases by the male, at 1427:15 he began an extended series of Chases, Extends, Twitches and one Shake (S), lasting until perhaps 1430:15. Again Extensions, Chases and Twitches were sporadically seen, interspersed with female Chases and Bites, until 1441:10 when the male began another long series of actions. Let us examine the first 30 seconds of minute 1442. The male Extended his paddle six times, then Twitched it several times before the female responded with Following and Nipping. At the second Nip he lowered his paddle and went into a typical Quivering (Q), pushing the female about the tank. She joined him, (Pairing, P) and they intertwined (Rolling, R) and leaped from the water.

Following their separation, the male continued to Chase and Extend, but was not observed to Twitch.

The black lines above the letters provide a means of recording changes in the belly spot color, which was observed to darken during courtship. As changes in

[5] In observations of this species Approaching, Moving Away, Dropping and most comfort movements were usually not recorded.

light intensity, background, and orientation of the male relative to the light and to the observer caused the belly spot to vary in appearance, it was not practical to utilize more than a black-not black dichotomy. Accordingly, the black lines indicate time during which the belly spot was black (it is at the top of the line for the corresponding minute).

Changes in frequency of activity with time.—Figure 14, *a* shows graphically the changes in frequency of selected social activities over a two-hour-long period. The bottom graph gives the number of seconds out of each minute that the male's belly spot was black.

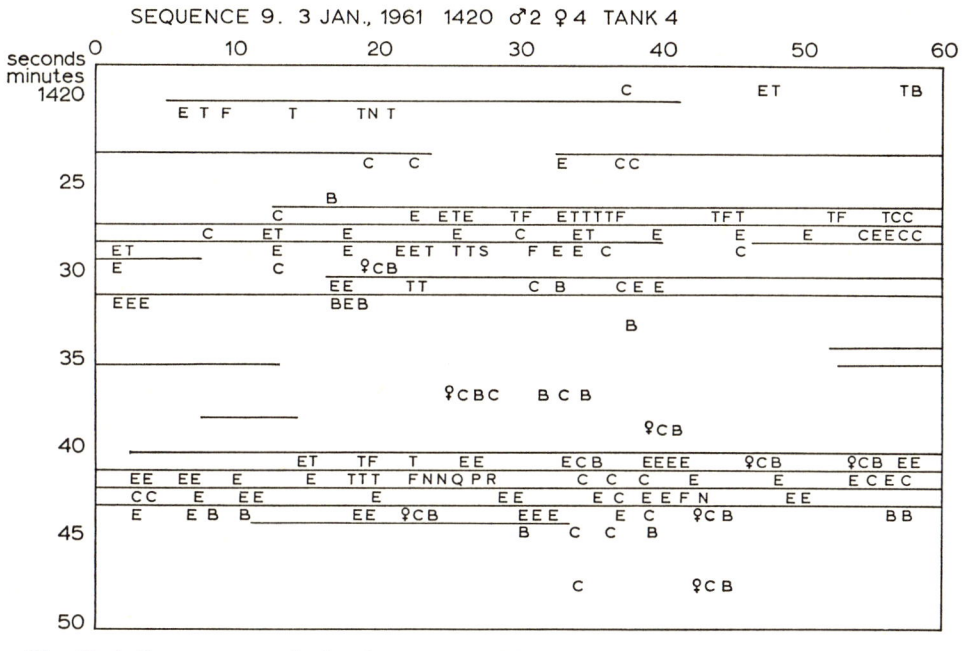

Fig. 13. A *Corynopoma* male-female encounter. The figure is to be read from left to right and from top to bottom (one line equals one minute). (Explanation in text.)

Notice that all activities except Upright Postures and perhaps Chasing, peaked at minutes 88–92, and that they tended to peak at other times as well, at about 20-minute intervals in this example.

These characteristics—a rise in activity at the start of the encounter, and sporadic peaks—were typical, although the rise at the beginning was usually much more rapid. If the male were left with a female for several days, a sort of equilibrium ensued, broken by sporadic courtship sequences.

With several females he would be likely to concentrate his activities on only one of them, for unknown reasons.

Records of selected activities of other males are given in figure 14, *b* and *c*. The apparent periodicity of peaks of activity did not survive a careful examination. By definition, at least one minute had to intervene between two peak minutes. When this factor was removed, and there were sufficient peaks to make the test, the distribution of minutes between peaks could not be shown to be different from a random (geometric) distribution (see Nelson, in press).

The pattern of courtship sequences.—A full description and statistical analysis of the temporal patterning of *Corynopoma* courtship is given elsewhere (Nelson, in press). The following brief summary is intended to indicate the general nature of that patterning, for comparison with the other glandulocaudine genera.

Periods of courtship could be separated from periods of non-courtship only in statistical terms, for if the frequency of intervals between pairs of adjacent male courtship acts was plotted against their durations, a smooth curve, showing no bimodality, resulted. A *sequence* of male courtship was defined as a series of statistically dependent acts separated from other such series by intervals separating statistically independent acts.

Sequences as defined nearly always began with either Chase or Extend, in the ratio 4:1. Within sequences no particular order of male courtship acts was seen to predominate; rather, practically any ordering was possible. The female responses Follow and Nip were rare, and most frequent following Twitch. Nipping following Twitching led into Quivering by the male more often than could be expected by chance, and the series Twitch-Nip-Quiver led much more often into Pairing and Rolling than did any other series (Nelson, in press). Rolling was almost always immediately followed by Chasing, and the continuation of the sequence. It was observed too infrequently to decide whether its performance led to a decrease in the probability of its own recurrence, but there was no evidence that this was the case. Certainly neither Quivering nor Rolling led to a decrease in the probability of occurrence of other male courtship activities.

Sequences were found to end randomly, that is, without regard for when they began. Sequences, as opposed to isolated courtship acts, were paralleled by the color of the male belly spot; up to 90 percent of male courtship acts occurred while the belly spot was black. From sequence to sequence, and even in some cases from day to day, the probabilities of occurrence of the various male courtship acts remained relatively constant; there was little indication of marked shifts in probabilities of occurrence during the course of a sequence. Thus, the courtship behavior of the male *Corynopoma* was characterized by relative statistical immobility.

COURTSHIP IN PSEUDOCORYNOPOMA

Description of courtship.—With *Ps. doriae* the encounter technique as a rule proved impractical because of the excessive shyness of the animals, which made observation difficult in any event. Instead, a pair was kept together for long periods of time while observations were being made of other species in adjacent tanks. During these observations the observer's attention and the operations recorder were switched to the *Pseudocorynopoma* tank at the first sign of courtship activity. In this way long records of nearly continuous courtship were obtained.

Whereas in *Corynopoma* individual events generally lasted less than a second, and measures of total activity were more meaningful in terms of frequencies, in the slowly moving *Pseudocorynopoma* durational measures were superior.

Typically the male would spend much time slowly Approaching (here considered as courtship) the female and then as slowly swimming away again. Occasionally,

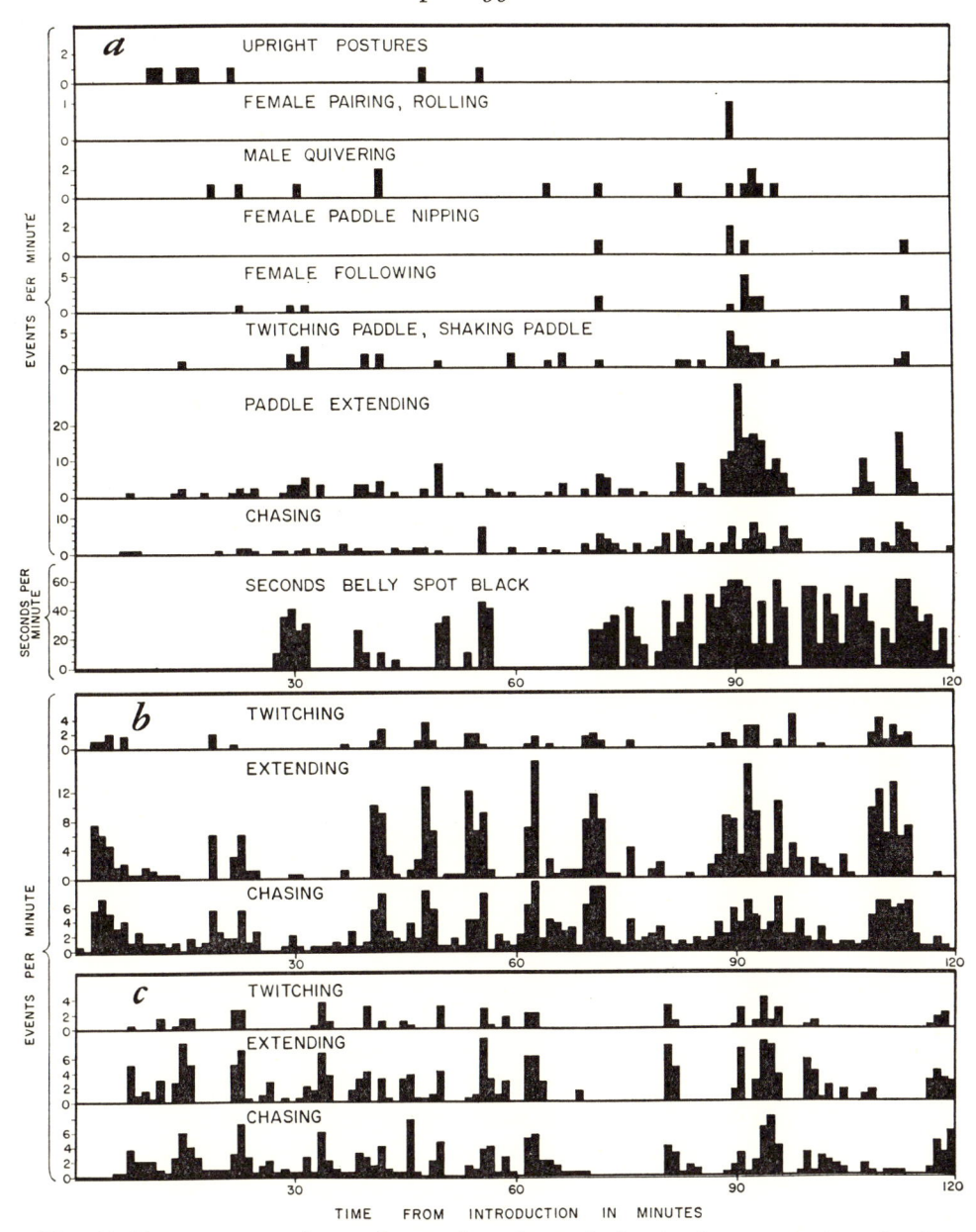

Fig. 14. Frequency per minute of a number of acts during two-hour encounters, for three males of *Corynopoma*: *a*. Male 2, January 8, 1961. *b*. Male C 2, October 24, 1961. *c*. Male J, November 1, 1961.

perhaps every ten minutes, these Approaches would become more frequent, and he would begin to Oval about her. Ovals were usually short to begin with and gradually increased in duration and tempo. At the same time that Ovals came to occupy about 40 percent of each minute, the time spent Approaching decreased to about 10 percent, and occasional, single upward Zigzags began to appear. As Zigzagging became more frequent, the percent of time spent Ovalling dropped, Approaching usually remaining at the same low level. If the female began to Follow, Zigzagging increased suddenly and markedly in intensity and as suddenly turned into Wobbling. When this occurred the female, if she had been Following, usually began Wobbling after the male, and with a sudden rush they would Impinge. After this the male would begin Ovalling again, or Approaching. If the female did not Follow, Zigzagging often continued to increase in frequency and duration, and as the direction of upward motion approached the horizontal, would merge into Wobbling. At this point the female, if she had not already done so, would often begin to Follow, with the result that Wobbling increased in intensity—that is, that the lateral undulations became more rapid. When this happened, the female would often begin to Wobble also, with Impinging as a result.

If the female neither Followed nor Wobbled, the male would usually continue to Wobble back and forth around the female, usually so that the plane of his body was perpendicular to a line from him to the female (fig. 7, *a*). As he was usually in a plane beneath her during Wobbling, the overhead light often flashed from his silvery body at her as he undulated, with striking effect, at least on the observer. Often this Wobbling would be continuous for over a minute, and then the male would either cease courtship altogether for a minute or so, or begin the slow Approaches anew.

Analysis.—Figure 15, *a* portrays graphically the changes which were observed during an hour's courtship of an unresponsive female of *Ps. doriae*. Zigzagging and Wobbling are here lumped, because in the presence of an unresponsive female, they merge imperceptibly. The first thing to be noticed is that courtship, if defined to include Approaching, occupied most of the animal's time during this record. In most minutes, from 50 to 70 percent of the record is of courtship. Secondly, the time per minute spent courting as a whole has a smaller variance than do the component acts; in about three-fourths of the minutes of the record of figure 15, *a* the animal spent between 30 and 50 seconds courting. Corollary to this, the three courtship categories listed are mutually negatively correlated ($r_s > .575$). This is in contrast to *Corynopoma*, in which the frequencies (not total durations) of courtship acts peaked together (see fig. 14).

Next, it will be seen from figure 15, *a* that Zigzag-Wobbling builds up gradually to a peak, and suddenly drops to zero (in the peaks of minutes 19, 27, and 54, the period of continuous Wobbling overlapped into the next minute). Wobbling is thus consummatory in that it is accompanied by a lowering of the probability of its own recurrence. In *Corynopoma*, on the other hand, the only evidence that an equivalent activity such as Shaking or Quivering is consummatory is that it leads into Chase more often than would be expected to occur by chance.

Thus, a sequence of *Ps. doriae* courtship, as measured from one bout of Wobbling to the next, has a structure which is considerably different from that in *Corynopoma*. The last sequence in figure 15, *a* is more typical than the rest, and may be diagrammed as in figure 15, *b*. There the probabilities that at any moment one would have found the male engaged in Approaching, Ovalling, or Zigzag-Wobbling in the absence of female response are graphed. These are seen to be

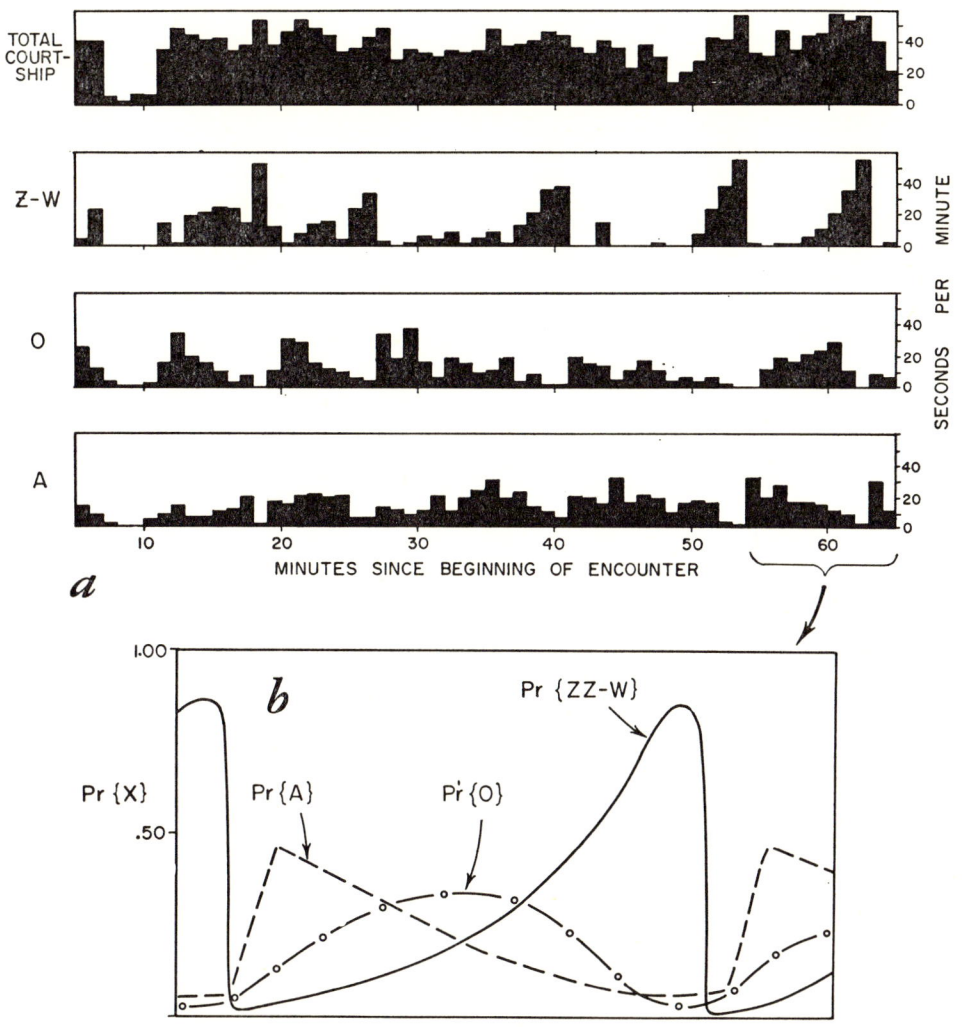

Fig. 15. *a.* The number of seconds per minute during which selected male courtship activities were seen in a *Pseudocorynopoma* courtship record. A, Approaching; O, Ovalling; Z-W, Zigzag-Wobbling; total courtship, the sum of the number of seconds in each minute during which the three acts occurred. Male 2, March 4, 1963. *b.* A schematic representation of the data of the last 12 minutes of the record in *a*. Pr [×] is roughly the probability that × will be seen at any instant. The abscissa represents the time enclosed by the bracket in *a*.

heavily dependent upon the time of last occurrence of Wobbling. In sequences with female responses the results are the same: a preliminary period of Approaching, then Ovalling; a gradual rise to a peak, Wobbling, Impinging, and then either Approaching or a period of no activity.

Female responses and their effect upon the behavior of the male are quite marked. More often than not, if the female Followed, a sudden increase in the intensity of Zigzagging was observed. Also, if the female was responsive, she tended to be so in an all-or-none fashion: if she Followed, she would probably Wobble; if she Wobbled, she would Impinge. These observations appear to hold equally well for *Ps. heterandria*.

Courtship in Coelurichthys

DESCRIPTION OF COURTSHIP PERIODS

C. microlepis.—A "pure" courtship encounter, without intermingled Lateral Displays, is difficult to obtain in this species. The closest thing to it was obtained using Male 4 who was much larger than any of the females. The following description is of one of his encounters, and is atypical in that Pairing and Rolling occurred after only 23 minutes, and Lateral Displays, Vibrating and Jerking were rare.

0907. Introduced male 4 into tank 3 with female A. He hides at rear; A patrols the central area. 0910. Male 4 follows A; he then Approaches again, A Lateral Displays, fins folded (LDff). Male 4 Zigzags at moderate speed, then speeds up as A Follows. Zigzags last for 2 minutes; then he begins Hovering over her, head down. 0916. He follows her about the tank, then Hovers above her, circling and Jerking. There follow several minutes of alternate Chasing and Hovering, then the swooping sort of Zigzag begins; when she Follows it increases in amplitude and frequency. Then a minute or so of Hovering. 0923. Male 4 Zigzags again, in tight circles about her, with the upward component at 75° to the horizontal. She moves off, he follows, Hovers, follows, she Approaches, he LDfu's momentarily. 0928. He alternates between slow Zigzagging and Hovering; then, as she moves slowly away, this turns into high frequency Zigzagging, which lasts for about 20 seconds. He stops, turns toward her, and they Pair, swim up to the surface in unison and swim in one or two circles, then Roll at about 0930. They separate and the male Chases and Hovers, then Chases, occasionally Jerking.

Usually courtship was much more prolonged. Typically Chasing, Hovering, and Zigzagging would succeed one another, usually with little in the way of female responses, for hours on end. Bout length in general appeared to depend upon the female; if she actively swam about, Chasing was more common, but not necessarily more prolonged, and bouts of Hovering and Zigzagging were shorter and more frequent. These long, continuous periods of courtship were seen only in well-established groups, except as in the above encounter with Male 4.

C. tenuis.—Initial encounters in this species invariably involved Lateral Displays on the part of one or both participants; as mentioned earlier, in both species the male had to establish his dominance before continuous courtship occurred. The following fairly typical encounter will indicate the nature of the transition.

1300. Placed Male 3 in tank with Female B; he Drops. 1302. B approaches, LDfu's, he LDfu's, they move Head-to-Head (HH). She follows him and he LDfu's continuously; she occasionally joins him, HH. 1306. He moves from LDfu into Hovering, rolls toward her; then he Chases at her, she Moves Away, then back to Hovering. 1308. He continues to alternately Hover and Chase; occasionally Chases are *at* her, as in Chase-Bite. 1312. He follows, Hovers, moves forward and Dusts a few times, then follows (Approaches), Chases her in circles, Hovers, Approaches. 1315. Chases her some more. All this time—except in Chases—his fins are moderately raised. More Dusting, Chasing. 1318. Hovering, Chasing, Hovering. Tilt during Hovering is more pronounced, so that the plane of his body is at a 45° angle. More Dusting, Chasing, Hovering; at 1324 he again Chases at her in short bursts and Bites; he moves off. 1325. He returns to Hovering and Chasing. This continues, with Dusting and occasional Chase-Bites until 1347, when during Dusting she Follows. More Hovering, Dusting, etc., with occasional

Zigzags, usually of miniature size. At 1349 Dusting is semi-continuous; then more Zigzags, and back to Dusting, etc., until 1355 when he Chases, Bites, she Drops. At 1356 he is back to Hovering, Chasing, Zigzags, and Dusting, which continue until 1424, when the following series occurs: C-Chasing in circles; one symbol to a circuit; H-Hovering; D-Dusting, Z-Zigzagging, one Dusting movement to a D; A-Approach; ACDDDDDHCDDDADDDDCDD Pair Roll C ZC ZHA Pair (short) ZAZDDDCCDDDDADD Pair etc. During Hovering the fins are moderately erect; during Dusting all but the caudal are folded. The Pairs, except for the first, are very short, almost as in HH except the fins are folded. Rolling occurs at 1425.30, others at 1430, 1436.30, 1446, 1448, 1449.30, 1459, 1517.30. Rolling in these cases is momentary and it is doubtful whether fertilization occurs.

This encounter has been given in detail because it illustrates so well the way in which, in *Coelurichthys*, the shift from aggressive behavior to courtship occurs by shortening sequences of the one until they become interpolated events within long sequences of the other, and that once Pairing and Rolling have occurred, they tend to recur over and over. It is possible, however, that had true fertilization occurred, these repetitive Rollings might not have been seen.

Species comparison.—The most noticeable differences between *C. tenuis* and *C. microlepis* are the absence of Zigzags of large amplitude in the one and of Dusting in the other. Other differences of perhaps lesser importance occur. *C. tenuis* is much more active, and in it Chasing, especially of the Chasing-in-circles sort, is much more prevalent. Bouts of all activities, with the exception of Dusting to a responsive female, are shorter in *C. tenuis*, and are more often interrupted by Chasing. Following in the one species is in response to Dusting, and in the other to Zigzagging. Frequency or intensity of Dusting did not appear to change markedly when the female began to Follow; in *C. microlepis*, Zigzagging showed a marked shift in response to Following. Although the above descriptions do not demonstrate it, Jerking and Vibrating, usually during Approach or Hover, were more frequent in *C. tenuis*.

Nevertheless, the similarities—in Hovering, Chasing, Pairing, and Rolling especially—were marked. Both species showed practically continuous courtship for long periods of time (this and other features of courtship were shared as well by *Glandulocauda*). Thus, defining a sequence was not difficult and did not present theoretical difficulties; either the male was courting or he was not, once dominance had been established.

A more extensive analysis of courtship in *C. tenuis* is given elsewhere (Nelson, in press). Pairing and Rolling were most often preceded in *C. tenuis* by Dusting, in *C. microlepis* by high intensity Zigzagging. In *C. tenuis*, analysis showed that probabilities of occurrence of the various male courtship acts were stationary during any one sequence, but shifted between sequences with the gradual decrease in frequency of Chasing during the course of an encounter. This decrease appeared to result from a decreasing tendency of the female to flee, and resulted in the increase in length of courtship sequences mentioned above. Aside from this difference, and the more continuous and well defined nature of the *Coelurichthys* courtship sequence, the *pattern* in this genus resembled that in *Corynopoma*. Again, there was no indication that Rolling or other courtship acts lowered the probability of its own recurrence, or that of other courtship acts.

MATE CONSTANCY IN C. MICROLEPIS

In an earlier section, it was mentioned that there were signs of individual recognition in aggressive encounters. This was almost baroquely evident in courtship. In tank 4, the courtship situation described on page 109, example 1, was seen over and over again. Male 2, the larger male, would always court Female C, the larger female, and Male 3 would always court Female B. There was no problem of individual recognition by the observer, for although the difference in size between C and 3 was very slight, and although during a Lateral Display 3 appeared larger than C, the high degree of dimorphism rendered mistakes unlikely. During the course of several months' observations, Male 3 grew more rapidly than the rest, so that the size relation shifted to:

$$2 > 3 > C > B > D.$$

Removal of Male 2 always resulted in Lateral Displays, Biting, and quasi-territoriality. Removal of C resulted in 2 Lateral Displaying to 3, who for a time continued to attempt to court B. Soon 3 and B had Dropped, however, and 2 swam back and forth through the tank, but did not court B or D. If, however, 3 was then removed, eventually 2 would begin to sporadically court B. If C were then replaced, 2 switched his attentions to her, and C in turn Lateral Displayed, Chased and Bit at B. These shifts could be produced over and over again.

In the situation in which 2 and 3 courted respectively C and B, it was noticed that 2 and C were generally considerably darker. During Zigzagging especially, the four animals would intermingle completely and confusingly; when the pairs separated, 2 was usually back with C, and 3 with B. If they came out of the tangle courting the "wrong" females, both males would be seen to stop Zigzagging, turn and swim about the tank until they approached their proper partners again, and then both would begin again to Hover or Zigzag. On one occasion during a tangle of Zigzagging C moved over to a corner near the bottom and 2 continued to Zigzag for perhaps 10 seconds. He stopped, began swimming about the tank, returned to female B and the still-Zigzagging Male 3, Bit at 3, then swam about the tank some more and upon approaching the place where C was remaining quietly, immediately began to Zigzag. At this Male 3, who had Dropped, began Zigzagging by B again.

These and similar observations convinced me that in this species there existed, at least among the males, a pronounced ability to differentiate between different individuals. The basis of this is unknown; it is hard to believe that it depends upon size and behavior alone. During these courtship sequences the females were to the observer *behaviorally* indistinguishable (except for dominance relations), and size differences between females were not marked.

The fact that when Female C either removed herself or was removed by the observer from the courtship situation, Male 2 did not court Female B except as it were accidentally, suggests that more is involved than the largest male fixing his attention upon the momentarily largest female and the smaller males in effect taking the leftovers. When Male 3 and Female C were removed, some time was required before 2 fixed his attentions on B. When 3 was reintroduced, 2 no longer courted B, but neither did 3.

This situation of mate constancy was also observed in larger groups of animals in a 30-gallon tank, where, however, individual recognition by the observer was difficult and the relationships could not be followed for long periods. Thus, the *extreme* constancy observed may have been an artifact of confinement in the smaller tank, where in any event the choice of a female to court was limited.

Suggestions of similar mate constancy were observed in *Corynopoma* males. Here it was more a case of a male devoting a preponderance, but not all, of his courtship to one female; attempts to explore the bases of this were unsuccessful.

In *C. tenuis*, when several males and several females were placed in a tank together and a reasonably stable grouping had been reached, individual males would often court different females for several minutes at a time. Sooner or later, however, a Chasing pair in passing by another courting male would distract him from the object of his attention and he would begin to Chase after the first pair. In a tank (50-gallon) containing six or seven males the Chasing group as it raced around the tank would strip males away from their females and soon all six or seven of them would be Chasing after a single female. This usually broke up into a bout of Lateral Displays by the whole group of males; and the original female and one of her Chasers—it was usually impossible to tell which one—would escape and that male would begin courting. For this reason long courtship sequences were rare in *C. tenuis* aggregates containing more than one male, but frequent Pairings and Rollings were nonetheless observed in these situations.

Schwab (1939) records for *Glandulocauda* apparently similar mate constancy to that of *C. microlepis*.

ORIENTATION OF THE HOVERING MALE

Males of the two species of *Coelurichthys* and of *Glandulocauda* were observed to Hover both above and below females. In all three the orientation below was more frequent with faster-moving females. In *Glandulocauda* whether the male oriented above or below depended also on other considerations, discussed in the following section. In all three species, when the male was orienting below the female his sagittal plane was vertical and his frontal plane horizontal or nearly so.

In *C. microlepis* and *G. inequalis* the male Hovering above the female pitches (fig. 8, *a*), so that his body axis is at an angle of up to 55° with the horizontal, head downward. His sagittal plane remains vertical or nearly so.

In *C. tenuis* on the other hand (fig. 8, *b*), the male when Hovering above or to one side of the female rolls so that his saggital plane is oblique with respect to the horizontal, but his body axis remains horizontal. In continuous courtship bouts this is often so pronounced that in Hovering directly above the female his sagittal plane will be horizontal; it is thus approximately perpendicular to a line to it from the female. Most often, however, the rolling is not this extreme.

In both species, as well as in other glandulocaudine and non-glandulocaudine characids, Lateral Displays are often performed with both pitching and rolling out of the vertical. This is especially pronounced in single Broadsides and Circling head-to-tail. Some of these orientation changes during Lateral Displays are seen in figure 6.

Finally, on occasion *C. tenuis* males were seen to pitch rather than roll while

Hovering above the female. This was usually followed immediately by a return to the horizontal, so that it had the same appearance as a miniature Zigzag with the downward component first.

COURTSHIP IN GLANDULOCAUDA

As in *Coelurichthys,* there is in this genus little difficulty in distinguishing sequences of courtship from other activity; courtship actions occur in long continuous series. A typical sequence consisted of periods of up to several minutes Hovering, with occasional Zigzags interspersed, alternating with similar periods of Chasing. Nearly always, a bout of Hovering more than a few seconds in length contained one or more Gulps, during which the male dashed to the surface, took a mouthful of air, and returned to Hovering. During Chasing, Gulps rarely occurred except at the very end of the bout, and were usually the signal that the bout was over and that the male would then Hover.

When a hydrophone was placed into the tank, it was found that Croaking occurred almost entirely during Hovering, except at the very beginning of a bout of Chasing. Further, when a bout of Hovering began, Croaking never occurred until immediately after the first Gulp; and thereafter during the bout of Hovering, if the male stopped Croaking he generally did not begin again until he had taken another Gulp.

Rolling was not observed in this species, Pairing was seen only once, and unfortunately the sequence which led up to it was not monitored with the hydrophone. It began with the male Hovering over the female, who remained near the bottom. Hovering was broken only by short bouts of Chasing. Every few seconds the male would Gulp and return to Hovering; sometimes his return would appear to startle the female and she would Move Away, whereupon Chasing occurred. During one long bout of Hovering the female began to rise slowly toward the surface, and the male began to Hover alternately above and below the female, each orientation lasting perhaps a second. The Hovers were now interspersed with single Zigzags; the male would occasionally circle in front of the female and then re-orient, Hovering above her. As they neared the surface, he circled once in front of her, and then lined up alongside her. They thereupon began Pairing, and after circling at the surface several times, separated without Rolling.

Schwab (1939) gives an account of courtship similar to the present one, except that he states that as the female approached the surface, the male swam in circles and elipses *beneath* her. While in the one case here observed, Hovering beneath was more prevalent than at other times, the regular alternations of orienting above and beneath the female were the more noticeable feature.

Schwab estimated that in the pairing he watched, 15–20 seconds were spent circling at the surface. He does not describe anything resembling Rolling, but does state that at the end of Pairing "It was almost as if the two fish fell away from one another." In the present example Pairing lasted for at most 5 seconds; hence, it is possible that Schwab witnessed true insemination.

Further analyses of the *Glandulocauda* courtship pattern, of the physical nature of Croaking, and of the role of Gulping in respiration and courtship are given elsewhere (Nelson, 1963 and in press). No respiratory significance could be found for Gulping. The pattern of Hovering and Chasing by the male was in general similar to that in *Coelurichthys;* a bout of either ended independently of when it began. Superimposed upon this random pattern, however, was the rhythmic repetition of Gulping and its associated Croaking during Hovering.

Zigzagging had in this species the appearance of an intention movement (Kortlandt, 1940) of Gulping, and occurred closer in time to a Gulp before the Gulp than after it. All gradations were seen between Zigzags, dashes toward the surface interrupted by pauses, and uninterrupted Gulping movements.

DISCUSSION

NON-SOCIAL BEHAVIOR

What has been called a "Beat" in the description of locomotory movements is a combination of Breder's (1926) "Carangiform" body movements and probably "Labriform" pectoral closing. The former are intermediate between the specialized "Anguilliform" and "Ostraciiform" movements, and are performed by the bulk of living fishes. Synchronous pectoral closing is termed "Arhythmic" by Wickler (1960), who considers it to be the generalized condition in living Characidae. The ability to swim in reverse and the presence of rhythmic pectoral movements in association with dorsal and upper caudal lobe vibrating in *Aphyocharax* and *Prionobrama* suggests a relationship to the Lebiasininae of Weitzman (1960) and indicates that their morphological and behavioral similarities to the Glandulocaudini may be a result of convergent evolution.

Hemigrammus, Moenkhausia, Bryconamericus, Knodus, Astyanax, and *Hemibrycon* are considered by Eigenmann (1917) to be part of the great central grouping of relatively generalized South American Characidae; Myers (1958) would consider them to be somewhat specialized, at least in terms of a reduction in dentition from the condition found in *Brycon*. In either event there is no indication that the Glandulocaudines have diverged significantly in locomotory adaptations, and this confirms the osteological conclusions of Weitzman (1962).

Weitzman (1960a) considers the morphological similarities of the carinate *Triportheus* and *Pseudocorynopoma* to be a result of convergent evolution. Their locomotory similarities may be considered to be also convergent.

In the Glandulocaudines the up-and-down component in feeding from suspended matter may be a mechanical necessity resulting from the oblique gape. However, the gape in young *Triportheus* is no more oblique than in many generalized characids, yet it also feeds upon suspended matter with an up-and-down movement. Thus, this component may be a behavioral carryover from the movements of surface feeding. The hovering way of taking food in *Aphyocharax* and *Prionobrama* again suggests affinities with the Lebiasininae. It should be emphasized, however, that this would be a taxonomically unorthodox grouping.

If, as appears likely, the Glandulocaudines subsist largely on floating food in nature, Nipping Surface and Nibbling should be properly classified as feeding behavior.[6] A strong positive correlation was seen in *Corynopoma* and *C. tenuis*

---

[6] In *C. tenuis*, Nipping Surface was without respiratory significance. Other species were not tested. (Nelson, 1963.)

between frequency of Nipping Surface and days since last feeding; this correlation was not found, however, in *C. microlepis* (Nelson, 1963). It is at this point that an attempted functional classification breaks down. With the exception of the catching and spitting activity found only in association with Nibbling, there was no discernible difference between the jaw movements of Nibbling and Chewing, nor between the fin movements associated with them. Furthermore, there was little difference in kind between Chewing and Gaping or between Fin-Flicking and Fin-Raising. As mentioned earlier, then, those movements which are seen outside a social context, and do not obviously subserve alimentation, locomotion, or reproduction, are here classified as comfort movements, *incerta sedis*. This procedure is being adapted because a strictly morphological classification was even more unwieldy.

Chewing appears akin to Baerends and Baerends' (1950) "Mumbling," and Gaping is equivalent to their "Yawning." It may be that both are given as a result of irritation in the mouth or gill-chamber,[7] or that the Gape is a stretching movement, a real yawn, in which case we have another example of the association in the vertebrates of the stretching of gill-arch and other musculature (see Dumpert, 1921). However, it is also possible that Chewing and Gaping produce potential or actual bilateral asymmetries which are overcome by the stabilizing effect of median fin erection. Baerends and Baerends (1950) interpreted the raised dorsal fins of cichlids resting on the bottom as keels acting to prevent capsizing. Further discussion of these comfort movements will be given in later sections.

A survey of the hobbyist literature indicates that the scattering of adhesive, demersal eggs is the generalized condition in the majority of Characidae; fine-leaved plants appear to be the preferred spawning substrate. The Glandulocaudines studied all differ in that: (a) eggs are placed rather than scattered, (b) broad-leaved plants are preferred. *Aphyocharax* differs from both the generalized condition and from the Glandulocaudines in scattering non-adhesive eggs (Innes, 1948). In Rio Grande do Sul, where the present stock of *Pseudocorynopoma doriae* came from, this species is said to scatter adhesive eggs over fine-leaved plants (A. V. Schultz, 1962). A further discussion of the pattern spawning will be given in a later section.

It was not possible to confirm Innes' (1948) assertions of territoriality and egg care by females of *Corynopoma rüsei*, and his interpretation is considered unlikely on the following grounds as well. The eggs in this species were distributed all about the tank (24 inches wide by 30 inches long), in fact were less confined to particular loci than in the other Glandulocaudines studied, and thus would have been singularly hard to defend successfully. It should be borne in mind, however, that continental populations of *Corynopoma* may exhibit the behavior which Innes describes. I consider it unlikely that the absence of the entire pattern of parental care was an artifact of captivity in the present observations.

AGGREGATION

The ontogeny of aggregation in the three species studied suggests that the young grow up in an environment with dense cover, and then move into more open water

---

[7] In correspondence, G. Barlow has remarked upon the fact that "coughing" and "backflushing" were not observed. I may have confounded these with Chewing.

where they find it advantageous to aggregate. This coincides with the ecological observations described in the section on Ecology and Distribution, and with the postulated evolution of the group given below. Morris (1958) compares the shoaling in young *Gasterosteus* with the scattering of the young of the related *Pygosteus* (= *Pungitius*) after they leave the "nursery." The latter nests in denser vegetation where individuals are perhaps better protected by scattering for cover than they would be by shoaling. The Glandulocaudines are similar in this respect to *Pungitius*, but, as they are not protected by spines, schooling without respect to cover is preserved in the adults.

In this connection several observations on aggressive behavior are interesting. In *Corynopoma* adults, Chasing and Biting were least frequent during aggregation, rose to a peak at its end, and recurred sporadically throughout the period of dispersal. In general, the other species were similar in these respects. These facts suggest several possibilities. Either animals will only tolerate the closeness of others for a limited time; or Chasing and Biting are somehow suppressed or inhibited by some factor related to aggregation, such as its initial stimulus; or aggregation simply reflects what the fish do when they are not Chasing or Biting. The peaking of these actions at the end of aggregation, together with the fact that they did occur sporadically during the period of dispersal, indicates that the second explanation is the more likely.

If aggregation upon approach of a possible predator provides a means of protection from predators (the "confusion effect:" Allee, 1951), then it is significant that Chases and Bites, which tend to lead to dispersion, are absent during aggregation; this is not simply a circular argument.

The Chasing of introduced juveniles by young fish, Chasing during feeding upon newly hatched *Artemia*, and other observations indicate that Chasing in juveniles may in many cases be an unspecific response to anything too small to be a predator, that is, to anything which could conceivably be food. Then, within limits, the distinction between food and not-food would be made for, rather than by, the young fish: if the Chased object be uncatchable then it is not food. However, if Chasing does lead to dispersion, and if aggregation is advantageous in adults but not in juveniles, then the increased Chasing in the latter observed during feeding may serve to prevent non-adaptive overaggregation.

Breder (1951) has considered the possible results of interactions between aggressive behavior and the tendency to aggregate in fishes. He hypothesizes that if the two balance, fish will be "socially neutral," whereas the prominence of the former will lead to a solitary life, and that of the latter to schooling. With high levels of both, but with the aggregating tendency predominant, a peck-order will result. "This is tantamount to saying that aggregating forms would all school if it were not for the dispersive influence of the aggressive behavior, which leads to the establishment of a measurable peck-order. Viewed in this way the 'fright school' is merely the suspension of the peck-order under the duress of a strong extroceptive influence" (Breder, 1951, p. 21). Breder's interpretation affords a good first approximation for glandulocaudine aggregating behavior. In *Corynopoma* groupings, the aggressive, peck-order-forming tendency is weakly developed, and correspondingly they aggregate upon less pronounced external stimulation.

However, there is evidence that more factors are operative in regulating aggregation. For example, upon strong stimulation quasi-territorial *C. microlepis* sometimes joined an aggregation but often did not, instead swimming rapidly about the tank, apparently at random. Often in stable groupings of *Corynopoma* females, repeated aggregation and dispersal was seen with neither evident external stimulation nor evident Chasing. The degree of dispersion of aggregations of *C. tenuis* fluctuated widely without corresponding fluctuations in aggressive behavior; conversely, in this species the level of aggressive activity often varied without corresponding changes in the degree of aggregation. Thus, in these species the observed degree of aggregation was seen to vary independently of aggressive activity, and there is no reason to think that the aggregating "tendency" did not itself increase upon external stimulation.

Individual Recognition

There was evidence that individual recognition played an important part in social interactions of all species studied. The establishment of dominance hierarchies in groups of *Corynopoma* males, *Glandulocauda*, and *C. microlepis* indicated that individuals in these groups could differentiate between one another. *Corynopoma* males, when offered a choice of females to court, showed preferences for particular females, not necessarily the largest nor the least aggressive. But the strongest evidence for individual recognition came from the experimental alterations of groups of *C. microlepis*, in which consistent differences in behavior toward different animals could be demonstrated to last over periods of several months. Indeed, the mate constancy exhibited by the males of this species was so marked as to suggest that in nature *C. microlepis* may exhibit something approaching a true pair bond, which, however, would be initiated and maintained only by the male. The maintenance of such a relationship in dry-season pools would appear to offer no difficulties. Its functional significance may be inferred from the aquarium observations that courtship, if it is ultimately successful, (a) is preceded by the establishment of male dominance, and (b) is continuous or nearly so over a long period of time. Thus, the restriction of a male's attentions to one particular female—no matter which one—would obviate the necessity for repeated reëstablishment of dominance, and would help to insure that courtship would be continuous.

In *C. tenuis* groups, males did not give their undivided attention to individual females, although the need for the establishment of dominance was evident in encounters between a single male and female. Pairings occurred with greater frequency in crowded tanks than was the case with *C. microlepis*. *C. tenuis* showed a higher tendency to aggregate, and a lower tendency for aggressive behavior to break up the aggregation than did *C. microlepis*. Thus it is possible that the solution to the need for long periods of continuous courtship preceding pairing has in this species involved group courtship under conditions of high aggregation. Ecologically, if similar dry-season pools could support an equivalent biomass of either species, a not unreasonable assumption, in *C. tenuis* this biomass would consist of many more individuals than in the larger *C. microlepis*. Hence, it may be that whereas *C. tenuis* can afford dry-season aggregation and group courtship,

C. microlepis cannot, and has instead chosen the mate-constancy route to successful fertilization.

Corynopoma riisei males exhibit a strong mate preference, but not mate constancy, and this species is correspondingly intermediate in size between the two species of *Coelurichthys*. Thus, although it is realized that the ecological and behavioral data from the field which are needed to substantiate the above inferences are lacking, it is felt that the necessity to persuade unresponsive females to mate which dominates other aspects of glandulocaudine behavior and morphology (see below) is again involved in the phenomenon of individual recognition displayed by male *C. microlepis* and *C. riisei*.

In groups of female *Corynopoma*, it was observed that the effect which the introduction of a newcomer made upon the behavior of the group was inversely related to the group's size. Thus, the addition of one female to a tank containing a single resident usually resulted in extensive Chasing and Biting, the addition of an individual to a group of two or three caused somewhat less change, whereas the introduction of a newcomer to a large group created little or no disturbance. The situation among *C. tenuis* of both sexes was similar, although interrelations were somewhat more complex. In *C. microlepis*, considerable disturbances were seen when an animal was added to groups of even six or seven individuals, but with larger groups the effects became less pronounced.

It may be that the inverse relation which was seen was itself connected with the relative change in size of the group created by the introduction of a newcomer. An animal added to a tank containing a single resident represents a 100 percent increase in group size; as the initial size of the group increases, the percent change created by the addition of a new animal rapidly diminishes as the reciprocal of group size. It is postulated that instability of group interactions is dependent in part upon apprehension, probably visual, of changes in group size. If this is so, individual recognition may play a role in modulating this effect. If a member of a group is capable of differentiating accurately between only a few individuals, he will always be able to detect the presence of a newcomer in a group of less than that number of individuals. The degree of disturbance upon addition of a newcomer to a group of *Corynopoma* females or *C. tenuis* followed fairly well the reciprocal relationship outlined above; it was postulated earlier that in these species the faculty of individual recognition was poorly developed. In *C. microlepis*, on the other hand, evidence was presented to indicate that the ability to differentiate between individuals was pronounced, and marked disturbances attributable to the presence of a newcomer were seen even in groups containing six or seven individuals. Finally, heterosexual groups of *Corynopoma* exhibited greater deviations from the expected reciprocal relationship than did all-female groups or mixed *C. tenuis* groups, and the evidence from the establishment of dominance hierarchies and courtship preferences indicated that in terms of individual recognition, male *Corynopoma* were intermediate between *C. tenuis* and *C. microlepis*.

The sensory bases for individual recognition were not explored systematically; however, certain inferences may be drawn from the results of interspecific encounters. On only one occasion was indisputable interspecific courtship seen; this occurred during a long *Corynopoma* courtship sequence, when a catfish (*Corydoras pestai*) passed behind the male, who Extended. Approaches, but no more, have been observed between species of *Coelurichthys*. Systematic observations of encounters between the two species of *Pseudocorynopoma* have not been made, but preliminary observations revealed no interspecific courtship.

On the other hand, interspecific aggressive behavior—Lateral Displays, Chases and Bites—were common. The large male *Glandulocauda*, which Lateral Displayed only rarely to members of its own species, was seen to engage in long bouts of Lateral Displays to a female *Corynopoma* much larger than itself. Male *Pseudocorynopoma* of both species exhibited aggressive behavior to both sexes of the other *Pseudocorynopoma* species, *Corynopoma* and *C. microlepis*, but with *Ps. doriae* at any rate, males of these species elicited more prolonged aggressive behavior than did females. The fact that *Ps. heterandria* were aggressive toward similarly sized *Bryconamericus*, *Knodus*, and *Hemibrycon* indicated that in this species the aggression was not released by specifically glandulocaudine characters, but the more pronounced *Ps. doriae* aggressive behavior toward male Glandulocaudines may have been elicited by their longer fins.

However, *C. microlepis* males and females showed more aggressive behavior toward *Corynopoma* females than toward males; to the observer *Corynopoma* females more closely resembled *Coelurichthys* in general morphology than did *Corynopoma* males. Thus, there is every indication that in the Glandulocaudines rather unspecific visual cues could have been important in releasing interspecific aggressive behavior.

On the other hand, it is not so clear that visual cues played an important role in individual recognition of females by males. In *Corynopoma*, for example, males often Displayed and even Quivered to other males (fig. 12), although long courtship sequences were not seen in male-male encounters; as mentioned, a male was once seen to Display to a catfish. Yet male *Corynopoma* were evidently able to distinguish between similarly-sized and colored females in courtship. Similarly, male *C. microlepis* were able to distinguish between females of only slightly different size and color. Also, interspecific aggression, but not courtship, was seen. It appears that although unspecific morphological characters were sufficient to elicit aggressive behavior and, in males of *Corynopoma*, isolated courtship acts, individual recognition and fully-developed courtship were dependent upon the reception of more specific features, probably of a behavioral and chemical as well as a morphological nature. A further discussion of the nature of stimuli involved in courtship will be found elsewhere (Nelson, in press).

Systematics and Phylogeny

The status of Coelurichthys, Mimagoniates, and Glandulocauda.—The history of these generic names has been discussed above. Whether or not their retention as used in the present paper is allowed to stand, the species subsumed under them appear to be quite distinctive. Contrary to the opinion of L. Schultz (1959), *Glandulocauda inequalis* and *Coelurichthys* (= *Mimagoniates*) *microlepis* are quite easily distinguishable. In addition to possessing two rows of anal hooks instead of one and much less modified caudal fin rays, males of *G. inequalis* differ from *C. microlepis* in having a straight instead of emarginate anal fin containing fewer rays, a virtually invisible lateral band, a more markedly convex dorsal profile, and a generally deeper body. Behaviorally, the male *C. microlepis* lacks the rhythmic Gulping and Croaking characteristic of *G. inequalis*, and has a more pronounced and quite distinctive Zigzagging movement.

Contrary to the opinion of Myers (Eigenmann and Myers, 1929), *C. tenuis* is quite distinct from *C.* (= *M.*) *microlepis*. As Nichols (1913) correctly observed, the lateral stripe in *C. tenuis* continues onto the chin, and the male's caudal tube points downward and backward at an angle, whereas in *C. microlepis* it is nearly horizontal. In addition, all the *C. tenuis* of the present study were "emaciated" when compared with *C. microlepis*. The colors of the two species, even in alcohol or formalin, are quite different, and H. Schultz (1959 and pers. comm.) mentions a habitat difference as well. Behaviorally, *C. tenuis*, while lacking the pronounced Zigzagging movements of *C. microlepis*, habitually courts with a movement, Dusting, which is rare and incomplete in *C. microlepis*.

M. barberi Regan is distinctive in possessing several rows of anal hooks, and lacking extensively modified caudal rays, as in *Glandulocauda*, but having an anal fin count much higher than either *Coelurichthys* or *Glandulocauda* (L. Schultz, 1959). Thus these characters should allow the most confirmed "lumper" to separate the four species to his satisfaction. In addition it has not yet been possible to demonstrate interspecific courtship in any of the *Coelurichthys* and *Glandulocauda* combinations which have been tried.

The ancestry of the Glandulocaudini.—As Eigenmann (1917) envisaged it, the family Characidae was composed of a great number of generalized species scarcely differentiated from one another, together with a number of small groups of divergent forms showing various degrees of specialization. He divided the great central group into two sub-families, the Tetragonopterinae (*Bryconamericus, Astyanax, Hemigrammus, Moenkhausia*, etc.) and the Cheirodontinae (*Cheirodon, Odontostilbe, Paragoniates, Aphyocharax*, etc.), recognizing that the distinction was probably in part artificial. He remarked (Eigenmann and Myers, 1929) that the Glandulocaudines appeared to be related on the one hand to such Tetragonopterins as *Bryconamericus*, and on the other to such Cheirodontins as *Paragoniates* (in habitus) and *Compsura* (by virtue of the specialized caudal scales).

Gregory and Conrad (1938) placed the glandulocaudine genera in their Characinae, composed in part of Eigenmann's Bryconinae, Characinae, Gasteropelecinae, Tetragonopterinae, and other subfamilies, and implied that the Glandulocaudines were derived from a form such as *Brycon*. Myers (1958), however, has emphasized that the multiple tooth row condition found in *Brycon* appears to be primitive for the Characidae, and the reduced dentition found in the Tetragonopterins and Cheirodontins appears to be a specialization accompanying their smaller size.

Böhlke (1954) again emphasized the resemblance of the "borderline" genera *Planaltina, Phenacobrycon*, and *Argopleura* to *Bryconamericus*, and considered a derivation of the Glandulocaudines from a form akin to *Brycon* as unlikely. As in dental formula and general morphology these "borderline" genera are scarcely distinct from certain *Bryconamericus*, I concur in Böhlke's opinion.

However, in having expanded suborbitals, posteriorly-placed dorsal fin and oblique gape, the Glandulocaudines resemble certain of the Cheirodontins, specifically *Aphyocharax* and *Prionobrama*, as well as *Bryconamericus*. Also, pairing above the water, typical of the Glandulocaudines, has been reported in *Aphyocharax rubripinnus* (Innes, 1948). Nevertheless, in dental formula, these Cheirodontins are quite distinct from *Bryconamericus* and the Glandulocaudines; and they exhibit rhythmic pectoral movements (Wickler, 1960) and often swim in reverse. The generalized Tetragonopterins examined (*Bryconamericus, Knodus, Hemigrammus, Astyanax,* etc.) and the Glandulocaudines uniformly show arhythmic pectoral movements and back up rarely and only with difficulty. Thus, derivation of the Glandulocaudines from a form such as *Bryconamericus* seems to require fewest assumptions, and the similarity of pairing in *Aphyocharax* to that in the Glandulocaudines appears to be a result of convergent evolution.

Weitzman (1960a), Böhlke (1954), and others emphasize that the Glandulocaudines are only slightly divergent from the characid central grouping. It appears that the morphological and behavioral differences which do exist are associated with the primary glandulocaudine specializations, top-feeding and internal fertilization.

The adaptations associated with top-feeding—oblique gape, large pectorals, etc., and the upward movement in taking suspended food—need little comment. Observations also show that in the Glandulocaudines Nipping Surface is much more frequent than in the more generalized characids. A possible role of Nipping Surface in the evolution of internal fertilization is discussed in later paragraphs.

The hypothesis of a polyphyletic origin.—Because the group as a whole is so little divergent from the generalized characid condition, doubts have arisen as to whether the Glandulocaudini form a natural group (Weitzman, 1960a). Böhlke (1958a) categorically denied that they are monophyletic. Although he has as yet not published a clear argument for this conviction, from his 1954 paper one may infer that his reasons are as follows. First, *Planaltina* and *Argopleura*, for example, are obviously close to *Bryconamericus*, yet they occur on opposite sides of the continent, and are therefore probably independent derivations from forms akin to the latter genus. Secondly, modified caudal scalation has arisen independently among the Cheirodontins, and thus may also have arisen several times among the Tetragonopterins. Third, Böhlke (1954) considers *Acrobrycon* to be "a specialized *Hemibrycon*," and indeed except for the caudal pouch it does more closely resemble *Hemibrycon* than it does other Glandulocaudines. Thus, if *Acrobrycon* is retained in the Glandulocaudini, only the presence of a caudal gland in the males distinguishes the group from the more generalized characids.

Böhlke's biogeographic evidence leads the present author to the opposite conclusion. The same disjunct distribution is exhibited by *Hysteronotus* among the Glandulocaudini and by *Compsura* and *Odontostilbe hastata* among the Cheirodontins. Other examples of this distribution among the Characidae could be cited. This suggests rather that the eastern Brazilian forms are relict populations of formerly widespread genera, and not independent derivations.

The comparative study of caudal morphology has led the present author to the conclusion that the unifying feature of the Glandulocaudines is not the presence of modified caudal scales, but rather the development of glandular areas around the middle caudal rays and the lower caudal fulcra. It will be argued in later paragraphs that these glandular areas have evolved in response to the peculiar requirements of reproduction in the Glandulocaudini, and that the modified caudal scales and fin rays are secondary developments accessory to the function of the glandular patches. Thus, the development of modified scales for whatever purpose in other groups may not be germane to the question of whether the Glandulocaudini are monophyletic.

The strongest evidence for a polyphyletic origin lies in the existence of *Acrobrycon*, which in terminal mouth, forward position of the dorsal, and dentition closely resembles *Hemibrycon*. However, not only does it possess pelvic, dorsal, and caudal fin ray hooks and glandular areas on both middle and lower caudal rays, but its "pouch" is closely similar to that of an undoubted Glandulocaudine, *Hysteronotus* (Böhlke, 1958a). I am convinced that the development of internal fertilization preceded that of the caudal gland, and the presence of caudal and even dorsal fin hooks in the male *Acrobrycon* certainly suggests that in that genus fertilization is internal (see the discussion below). We must then ask whether a *Hemibrycon* would be likely to develop internal fertilization. *Acrobrycon* and *Hemibrycon* are adapted to living in swift streams and do not show adaptations to top-feeding. As is shown in the next section, nearly all known cases of internal fertilization in fresh-water fishes involve groups that live near and feed from the surface in relatively quiet water. An argument as to why this is so is presented there. There is no known case of internal fertilization developing in fishes adapted to swift streams. If it is argued that the Glandulocaudine-like features of sexual dimorphism in *Acrobrycon* are not associated with internal fertilization, then it must be asked what basis there is for detailed convergence to the glandulocaudine condition. Thus it is likely that *Acrobrycon* developed its caudal gland and hooks prior to moving into fast-running waters, and once developed, these features and possibly internal fertilization as well were retained. If this is so, then its similarities to *Hemibrycon* may be due to convergence, and no difficulty remains in including *Acrobrycon* among the Glandulocaudini. In brief, to postulate the development of *Hemibrycon*-like habitus in response to a new environment requires fewer assumptions than to postulate detailed convergence to the Glandulocaudines in features of sexual dimorphism in response to some unknown feature of a quite different habitat.

The nominal Cherodontins *Compsura* and *Odontostilbe hastata* present a more difficult problem. I was only able to examine two males of *Compsura gorgonae*, and they proved to differ in a number of details. One, the smaller, exhibited hooks on the caudal rays and a suggestion of glandular development between caudal rays 10 and 11. The larger lacked these, but possessed strong interhemal spines, characteristic of male dimorphism in the genus *Cheirodon*. In other respects the two closely resembled each other and members of the genera *Cheirodon* and *Odontostilbe*. Only the smaller specimen bore resemblances to the Glandulocaudini, and there appear to be three alternative explanations for this.

The first is that it is a true Glandulocaudine convergent to the Cheirodontins. The evidence in favor of this, namely, caudal hooks and a suggestion of a glandular area, is much slimmer than for *Acrobrycon* considered above; in addition *Compsura* would appear to occupy a "glandulocaudine" habitat, and the argument against independent development of Glandulocaudine-like caudal morphology is thereby weakened. Another possibility is that the Cheirodontins are all derived from the Glandulocaudini. In support of this is the top-feeding habitus of *Prionobrama* and other Cheirodontins, and a moderate interhemal development and tendency toward a single premaxillary tooth row (diagnostic of the Cheirodontins) in *Coelurichthys*. However, distributional evidence (Eigenmann, 1915) is against this hypothesis, and it would appear to create far more problems than it solves. Finally, Böhlke (1954) and others have suggested that the superficial resemblances of these Cheirodontins to the Glandulocaudines are a result of convergent evolution. The evidence for this view, while stronger than in the case of *Acrobrycon*, is not completely convincing, and the verification of this hypothesis must await osteological or behavioral investigation.

The behavioral evidence that the Glandulocaudines are indeed closely related is as follows. All genera studied exhibited Zigzagging in one form or another, although in *Corynopoma* and *Glandulocauda* it was only weakly developed. Up-and-down movements during courtship are not mentioned in the literature of other characids. Pairing and Rolling in *Coelurichthys* and *Corynopoma* were similar and did not resemble spawning described for other characids. *Pseudocorynopoma* did not exhibit Pairing and Rolling, but its relationship to *Corynopoma* and *Coelurichthys* is on other grounds indisputable.

The up-and-down movement of glandulocaudine feeding was not characteristic of the generalized characids studied nor of *Aphyocharax* and *Prionobrama*. Nipping Surface in the absence of food was common only in *Aphyocharax* of the non-Glandulocaudines. Finally, spawning was similar in *Coelurichthys*, *Glandulocauda* and *Corynopoma*, and involved placing rather than scattering eggs. Careful placement of spawn is found elsewhere in certain territorial African characids, and in highly specialized forms such as *Pyrrhulina*, but is not recorded for the generalized characids in the literature surveyed. Finally, the very rarity of internal fertilization itself among fresh-water fishes should not be ignored when deciding whether forms which exhibit it are closely related.

It is concluded that the glandulocaudine genera as listed by Böhlke (1958a), plus *Coelurichthys*, have a common ancestor. This common ancestor was probably internally fertilizing, possessed glandular tissue at the base of the tail, lived and fed at the surface, and was not far removed from members of the present-day tetragonopterine genus *Bryconamericus*.

The Evolution of Internal Fertilization

The occurrence of internal fertilization among fresh-water Teleosts.—It must be emphasized that the following discussion concerns *only* fresh-water fishes; further, it concerns *only* internal fertilization. The evolution of viviparity, while certainly interesting, concerns us here only insofar as viviparity must presumably be preceded by internal fertilization. To my knowledge, the evolution of internal fertilization *per se* has thus far received very little attention.

Internal fertilization has been reported in the literature for the following groups of fresh-water fishes (Berg, 1947; Bertin, 1958; Innes, 1948; Nikolskii, 1961; Weber end De Beaufort, 1922):

Clupeiformes (Isospondyli)
　Pantodontidae
Cypriniformes (Ostariophysi)
　Characidae: *Creagutus*, Glandulocaudini
　Cyprinidae: *Barbus vivipara*
　Auchenipteridae: *Trachycorystes*
Beloniformes
　Hemiramphidae
Cyprinodontiformes
　Adrianichthyidae: *Xenopoecilus poptae*
　Goodeidae
　Jenynsiidae
　Anablepidae
　Poeciliidae
　Tomeuridae
　Horaichthyidae
Phallostethiformes
　Phallostethidae
　Neostethidae
Perciformes
　Embiotocidae: *Hysterocarpus*
　Comephoridae

Of these groups, all but *Hysterocarpus* and *Comephorus* are tropical or subtropical.[8] *Hysterocarpus* is the sole fresh-water representative of the viviparous surf-perches, and need not concern us further. *Comephorus* is a pelagic cottoid which has evolved viviparity endemically in Lake Baikal, the deepest and probably oldest body of fresh water in the world (Brooks, 1950). Development of viviparity in *Comephorus* is probably associated with the development of a pelagic existence in a form with demersal eggs.

Among the tropical forms, all but *Creagutus*, *Barbus vivipara*, and *Trachycorystes* are top-feeders, possessing characteristic adaptations toward surface life. These adaptations are lacking in certain vegetarian poeciliids and in *Acrobrycon*, but there is little question that the former were evolved from top-feeding forms, and the case of *Acrobrycon* has been extensively discussed above. The assertion that *Barbus vivipara* is viviparous has been rejected (Barnard, 1941) and thus the only real exceptions to the rule appear to be *Creagutus* and *Trachychorystes*.

Of the tropical groups, discounting *Barbus vivipara*, the Goodeidae, Jenynsiidae, Anablepidae, Poeciliidae, and fresh water Hemiramphidae are viviparous; all the rest lay fertilized eggs. Contrary to popular belief, the most extreme gonopodial development occurs not in the viviparous Poeciliidae but in the oviparous—and remarkably convergent—Tomeuridae, Horaichthyidae, and Phallostethiformes. This fact, and the remarkable radiation in sexual dimorphism among the Glandulocaudini, indicate that in tropical habitats with well marked wet and dry seasons,

---

[8] Recently Mees (1961) described a new fish, *Lepidogalaxias salamandroides* (Isospondyli, Galaxiidae) from temperate Western Australia, the males of which possess a complex structure which may be an intromittent organ. While *Lepidogalaxias* is neither tropical nor a surface feeder, it is related to tropical, surface-feeding forms which are poorly known.

internal fertilization is an adaptation of major significance by itself and not simply a minor acquisition preliminary to the development of viviparity.

Ecological correlates of internal fertilization.—In the Glandulocaudini for which information is available, in many other characids (Pickford and Atz, 1957, p. 238), and in *Horaichthys* (Kulkarni, 1940) and other tropical fish (Pickford and Atz, 1957) spawning occurs after the beginning of the rainy season. In the habitats of the tropical Glandulocaudines and *Horaichthys*, the available water is much reduced toward the end of the dry season, and the beginning of the rainy season is marked with floods. Wide areas of swamp, inundated grassland, and small temporary creeks and ponds, not otherwise occupied by fish, become available, and it is there that spawning probably occurs (H. Schultz, 1959; Price, pers. comm.; Kulkarni, 1940; and pers. obs.). In sub-tropical southeastern Brazil, the floods come at the end of winter, but otherwise the picture is identical (A. V. Schultz, 1962).

During the dry season, the parents of the next generation of all species including predators are concentrated in a fraction of the volume of water occupied at the height of the rainy season. The probability of finding a mate is thus very high, but conditions for the successful production of young are not good. The concentration of predators is high, that of food organisms probably low. In addition, in many areas the only water remaining near the end of the dry season is in shrunken ponds and mudholes in stream beds (see Beebe, 1945, for a description of these conditions) where the concentration of carbon dioxide is high and that of oxygen is low.

On the other hand, during the annual floods the area of water available for occupation is vastly increased. These newly inundated areas would seem ideal for the raising of young. A large and relatively unutilized supply of rotifers and small crustaceans feeding on rotting terrestrial vegetation is available. Because of shallowness and difficulty of access of much of the inundated area, predators are probably much less abundant there than in the permanent streams. But, as the number of adults per unit area will be much smaller in these newly submerged areas than in the permanent basins from whence they came, the probability of finding a mate will be much lower. If, however, adults can *mate* during the dry season (when the probability of meeting conspecifics is high) and then *spawn* during the annual floods in the newly submerged areas (with an unutilized food supply and freedom from predators), they will reap the benefits of both worlds, and suffer less from their disadvantages. Internal fertilization is a way of accomplishing this.

The adaptations of surface feeders.—It stands to reason that the initial colonizers of such newly inundated areas will be top-feeders, as they occupy the superficial strata which will overflow, and they are independent of substrate for a food supply (see fig. 16). In the Glandulocaudini and possibly in other groups, the top-feeding habit may have facilitated the development of internal fertilization and successful utilization of both environments in another way. In a later section the fact that in the Glandulocaudines movement to the surface plays a large role both in early courtship and in pairing will be discussed, and it will be suggested that these movements are in part derived from feeding movements, as in Nipping Surface. Pairing out of the water has been developed independently at least three times in the Characidae: in the Glandulocaudines, in *Aphyocharax* (Innes, 1948) and in the famous *Copeina arnoldi*, which spawns on overhanging leaves and then splashes the eggs with water until they hatch (Innes, 1948). All three are adapted to a life near the surface, and all three take floating food (pers. obs.).

In pairing above the water, as in *Aphyocharax* and presumably the ancestral Glandulocaudine, eggs and sperm are *initially* confined to the thin film of water which envelops the pair, whereas in species which pair beneath the surface, the sperm are free to move outward in all directions. Thus in the former case, the probability that sperm will chance upon eggs *immediately* upon extrusion is much higher, and it is proposed that this is its adaptive significance in *Aphyocharax*.

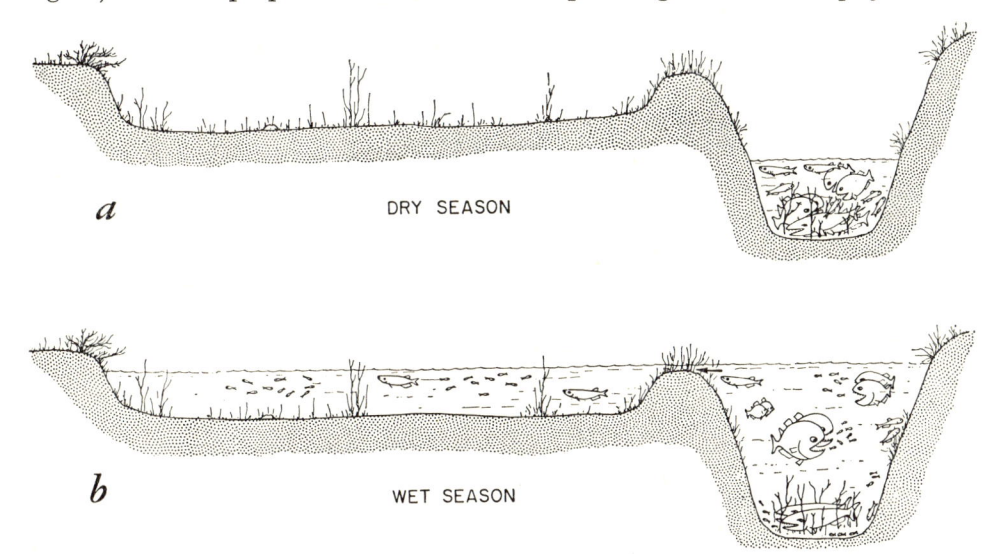

Fig. 16. The advantages of internal fertilization in a tropical habitat with well-marked wet and dry seasons. *a*. During the dry season the water level is low, and adjacent flood-plains and ponds are completely dry. Shown are Glandulocaudines near the surface, and various omnivores and predators below them. The probability of finding a mate is high. *b*. With the annual floods, the adjacent areas are flooded, and an abundance of food is available. The predators for the most part remain in the permanent stream; the Glandulocaudines, by virtue of their top-living, are the most likely to move into the new habitat, and there spawn successfully.

However, once the sperm are confined to the thin film enveloping the emergent pair, the probability is also much higher that they will find their way into the urogenital sinus of the female, and hence into her reproductive tract. Thus, it is proposed that pairing above the water, as found in fish with the top-feeding habit, facilitated the development of internal fertilization in the Glandulocaudini. In a later section it will be argued that although male Glandulocaudines have not developed the intromittent organs characteristic not only of most internally fertilizing teleosts but of selachians and most tetrapods as well, much of their sexual dimorphism can best be interpreted as "mechanical" adaptations increasing the probability of successful fertilization.

Behavioral and physiological correlates of internal fertilization.—Behavioral and physiological changes may be expected to accompany the temporal separation of mating and spawning as a result of the development of internal fertilization.

In externally fertilizing organisms it is essential that sperm and eggs be released simultaneously, and in such fishes as the Cichlidae (Baerends and Baerends–van Roon, 1950) and Gasterosteidae (for example, Morris, 1958) we see elaborate mutual courtship patterns to insure this. Although in the generalized Characidae the male is usually the more active, the female is by no means a passive participant in courtship, and sexual dimorphism is pronounced in only a few species (pers. obs., and aquarium literature). In many tropical characids, spawning is associated with the annual rise in waters, and rainfall, the rise in water itself, and temperature changes have all been implicated as triggers (Pickford and Atz, 1957, p. 238).

The problems confronting internally fertilizing species are somewhat different, and are different for each sex. As in externally fertilizing species, the female must remain responsive to environmental cues signalling optimal conditions for the early growth of the young (relative freedom from this necessity may be one of the major advantages of the development of viviparity, however). But if she remains oviparous, and is not to be continuously gravid, there should still be an environmental trigger releasing the train of physiological events leading to spawning. However, if there is now temporal dissociation between the behavioral event of mating and the behavioral and physiological events of spawning, it would appear that they must also become physiologically dissociated, that is, the female should have the behavioral potential for mating when she is not physiologically ready to spawn.

The problems facing the male are somewhat different. As the female can store sperm for long periods (up to seven months in *Corynopoma*; pers. obs.), there will be an adaptive advantage in continuous presence of both mature sperm and readiness to mate. Thus the common dependence of both on the same physiological system may be maintained, but there is no longer a need for, and indeed there may be a distinct disadvantage in, retaining dependence of this system upon specific environmental stimuli.

Many Brazilian externally fertilizing characids are difficult to spawn in aquaria and artificial ponds even at the height of the rainy season; the Brazilians have been preëminent in the use of pituitary extracts to overcome this problem (Pickford and Atz, 1957, p. 248). It is likely that in male Glandulocaudines, the pituitary-gonad axis has been freed from control by the rainy-season stimuli, but that while the pituitary-gonad-spawning relation to environmental stimuli has been retained in the female, her mating behavior has been freed not only from these environmental stimuli, but also from the old pituitary-gonad control.

The solution to these problems appears to have been quite similar in both the Glandulocaudines and the poeciliids (Clark and Aronson, 1951; Baerends et al., 1955). In both the female is rather unresponsive, and much of her responsiveness consists of passivity. On the other hand, in the tropical members of both groups the male spends much of his time throughout the year engaged in elaborate courtship activity; however certain poeciliids—those with large gonopodia—have relatively simple courtship patterns (Rosen and Tucker, 1961). In both groups, fertilization is rarely witnessed (pers. obs., and Clark and Aronson, 1951). This

relative passivity on the part of the female, and the compensatory extreme activity by the male, are apparently a result, at least in part, of the temporal dissociation of the physiological bases of mating and spawning in the female. It is postulated that while the *potentiality* for mating in the female has been more or less freed from dependence upon cyclic changes in the gonad-pituitary system, this has been achieved at the cost of *readiness* to mate.

Morphological and behavioral correlates.—Much of the evolutionary ingenuity in the Glandulocaudines has been expended in developing elaborate devices in the males for inducing non-responsive females to mate. The caudal gland appears to be such a device; it will be suggested below that it secretes a water-borne substance which raises the probability of female pairing. As will be demonstrated, the various accessory tubes, pouches and furrows in the different genera can best be explained as aids in the releasing, directing and transmitting of this substance. To this end, tubes have been developed independently at least twice, from the caudal rays in *Coelurichthys* and from one or more modified scales in *Pseudocorynopoma* and *Landonia*. A pouch is developed in all genera but *Pseudocorynopoma* and *Landonia*, but is reduced in *Mimagoniates, Coelurichthys*, and *Glandulocauda*. The caudal fulcra form a trough protruding from this pouch at its lower edge in at least *Corynopoma, Acrobrycon*, and *Gephyrocharax;* a trough formed of a modified scale protrudes from the upper part of the pouch in *Hysteronotus*.

Visual aids are present in *Corynopoma*, in which the male dangles a bait in front of the female, and probably also in such genera as *Pseudocorynopoma* and *Coelurichthys*, in which Zigzagging elicits responses from the female. However, the Twitching or Shaking paddle and body in *Corynopoma*, and the undulating body in *Pseudocorynopoma* and *Coelurichthys*, may also be involved in lateral-line stimulation and may in addition aid in transmission of the postulated caudal gland substance.

In *Glandulocauda*, Croaking by the male appears to be an inducement to Pairing; correspondingly the caudal gland, fin ray tube, and pouch are reduced from the condition found in *Coelurichthys;* strong Zigzagging or Dusting is not evident. It would be interesting to know whether *M. barberi*, which also has a reduced caudal apparatus, exhibits Croaking.

Summary.—In conclusion, the following evolution is postulated for the Glandulocaudines. It is proposed that the adaptations to surface feeding were the first to occur. When the prototype found itself in an area of strong annual floods alternating with a marked dry season, it was among the first to colonize the newly inundated adjacent areas. Those in which the development of internal fertilization and sperm storage could create a temporal separation between mating and spawning were at an advantage in utilizing the new habitat. This development was facilitated by the habit of leaping out of the water to pair. Once internal fertilization had evolved, subsequent adaptive radiation in the Glandulocaudines consisted largely of the development of new and divergent ways of insuring successful mating. This took the form of a flowering of sexual dimorphism, which may be unequalled in ingenuity and diversity in any other teleost group of similar age.

TEMPORAL PATTERNING IN GLANDULOCAUDINE BEHAVIOR

As in other respects (Tinbergen, 1951), behavior in its temporal aspects can be viewed as organized into a hierarchical series of levels of organization in time. The unit of time appropriate to each level, the number of distinguishable levels, the relationships between levels, and indeed the degree to which they are abstractions from reality should be expected to vary from species to species and even between different sorts of behavior within a species. Viewing the Glandulocaudines as a whole, three such levels appear to have particular biological significance. These are the annual cycle of behavior in nature, the sequence of behavior of a particular kind, and the behavioral act or event. Three intermediate levels may be named: cyclical changes of several weeks' duration, the daily cycle, and the bout consisting of acts of the same type. The first of these was difficult to define, the daily cycle was not studied, and with several exceptions, the bout appeared in the Glandulocaudines to be largely an abstraction. It may be observed that an annual cycle is a period of time about thirty millon times as large as a behavioral act lasting a second. With the exception of the daily cycle, each successive level represents a span of time approximately a power of ten larger than the previous one. Between a sequence and a daily cycle lie two to three orders of magnitude of time, and this is a convenient place to divide the temporal hierarchy. The larger temporal patterns will be considered first.

Long-term changes.—In the aquarium, there was no evidence of an annual cycle in courtship activities in *Corynopoma, Coelurichthys,* and *Glandulocauda.* Males were seen to court intensively in very crowded aquaria, at all pH values from 6.0 to 7.2, at temperatures approaching the normal annual range in nature of each species (exceeding it for *Corynopoma,* and at various degrees of water hardness. *Glandulocauda* males courted under conditions of high carbon dioxide and low oxygen (Nelson, 1963). On several occasions, *Corynopoma* males were dimly seen through an algal bloom to be courting intensely. At least for *Corynopoma,* the conditions they might be expected to meet in nature were not more extreme than those under which they courted in captivity, and there is no reason to suppose that in the wild the males do not court all year round.

On the other hand, although sporadic courtship was seen in *Pseudocorynopoma* during the summer, fall, and winter, it only became intense in the (Northern Hemisphere) spring. This is based upon observations of only a few males, but at least indicates that a courtship seasonality may exist in nature for *Ps. doriae.*

Corynopoma riisei is the only species for which sufficient aquarium spawning data are available, and in this species spawning was seen at all times of the year. There was no evidence for spawning seasonality in *Coelurichthys microlepis;* for *C. tenuis* and *G. inequalis* the data are too poor even to hazard a guess. Neither species of *Pseudocorynopoma* was spawned. The evidence from the field, however, indicates strongly that for all species studied but *Ps. heterandria* (for which no information is available), spawning occurs predominantly or entirely at the beginning of the rainy season.

Internal fertilization occurs indisputably in *Corynopoma, Glandulocauda,* and

Coelurichthys; for *Pseudocorynopoma,* reports are conflicting. Sterba (1959) and other German authors indicate that *Ps. doriae* exhibits internal fertilization, Innes (1948) states that fertilization is external, but mentions the other possibility, and A. V. Schultz (1962) indicates that it spawns after the fashion of other characids, the pair scattering eggs. *Pseudocorynopoma* was the only genus studied which did not exhibit Pairing at the surface; Rolling was seen only in *Coelurichthys* and *Corynopoma.* The present author is inclined to concur with Innes and A. V. Schultz, but the tantalizing possibility remains that internal fertilization in *Ps. doriae* is either facultative or subject to geographic variation.

The subtropical to temperate habitat of *Ps. doriae* and *Glandulocauda* exhibits more striking variations in temperature than do the tropical habitats of the other species. The season of annual floods in Rio Grande do Sul is preceded by a cold period, and it is possible that at that time the water level is higher than during the tropical dry season. Also, the winter cold may render the males too sluggish to court. Thus, I suggest that seasonal variations in temperature play a greater role in the breeding biology of the subtropical to temperate Glandulocaudines. Further, the stimulus of increasing day length is available to herald the onset of the warm annual floods.

In only two temperate fresh water genera of fishes, *Hysterocarpus* and *Comophorus,* has internal fertilization been developed. These and such subtropical poeciliids as *Heterandria* and *Gambusia* which venture into temperate waters are viviparous. I suggest that in the absence of the tropical dry season and in the presence of strong annual temperature and photoperiod variation, internal fertilization without viviparity is at best only a marginal adaptation, and may be actually disadvantageous. Further, the indisputably glandulocaudine *Ps. doriae* may as a result have lost not only the capacity for internal fertilization, but also such of its behavioral and morphological concomitants as Pairing, Rolling, and pelvic and caudal hooks.

On this view, *Ps. heterandria,* if it should prove to be externally fertilizing, would be a form derived from *Ps. doriae* which had reinvaded the tropics. Morphologically, in *Ps. heterandria* the ventral keel is more pronounced but sexual dimorphism is less extreme than in *Ps. doriae,* both of which differences are compatible with this hypothesis.

In summary, those changes which take place during the annual cycle would appear to affect the males of the internally fertilizing forms relatively little, except insofar as they raise or lower the probability of a male finding a mate, whereas the males of externally fertilizing forms, and females of all species, would be expected to be profoundly affected.

In *Corynopoma,* it was observed that the tendency of a male to court waxed and waned irregularly over periods of several weeks. In one male, the total number of courtship sequences per encounter was found to be highly correlated with the serial order of observations (Nelson, in press). As other males observed at the same time showed no such correlations, this may not have been dependent upon some such factor as increasing day length. Further, certain other courtship variables showed no correlation with the serial order of observations, but nevertheless displayed significantly too few runs,[9] demonstrating that fluctuations in those variables were not independent (Nelson, in press). Thus there was every indication that several factors underlying various aspects of male courtship waxed and waned over extended periods, perhaps quite independently of one another. Possibly these fluctuations were connected with successive waves of sperm maturation.

Spawning occurred predominantly in the morning. A survey of the aquarium literature indicates that this is usual among the externally fertilizing Characidae and in other groups as well. As hatching occurred approximately a day later, it also was predominant in the morning. Beyond this, there was little evidence of changes in behavior during the day, except for the absence of activity in the dark. With *C. microlepis*, there was an increase in courtship when sunlight directly struck the tank; in *Corynopoma* a similar phenomenon was sometimes seen, but equally often direct sunlight appeared to have an inhibitory effect. Intense light appeared to inhibit courtship in *Glandulocauda* and *C. tenuis*. Both courtship and aggressive behavior, apparently no different from that seen during the day, was seen in *Corynopoma*, *Coelurichthys*, and *Pseudocorynopoma* as late as midnight in artificially lighted tanks.

Thus, aside from the permissive role of light and the confinement of spawning to the morning hours, the daily cycle appeared to play little role in the behavior of the Glandulocaudines.

Finally, phenomena which appeared to be unconnected with the daily cycle, but which occurred at about the same temporal level or organization, may be discussed.

It has been previously emphasized, and will be further commented upon in the following section, that in an internally fertilizing species persistence of the male in courtship is a valuable character. Possibly associated with this in *Corynopoma* was the carry-over from one encounter to the next of the male's previous activity. Thus, males which had several hours previously been courting females vigorously would display more courtship activity when placed with males than would males which had previously been engaged in aggressive activity. Further, a male which Dropped in a preceding encounter was more likely to Drop when placed with a new male than was one which had been a Biter (table 3); it was possible to reverse these relationships by appropriate "training." This persistence or inertia of tendency resembled strongly what Tavolga (1956) has termed "set" in males of *Bathygobius*, and was not evident in *C. microlepis*. Once dominance relations between two individuals of the latter species were established in an initial encounter, they remained relatively constant, and the predicted behavior in an encounter was not altered by preceding experience.

This difference between *C. microlepis* and *Corynopoma* may be related to two other differences between the species. On the one hand, the capacity for individual recognition appeared greater in *C. microlepis*, and the differential behavior shown to different individuals persisted for months. In *Corynopoma*, although a male might show a preference for a particular female for several hours, this preference often shifted from female to female unpredictably. On the other hand,

---

[9] A "run" being a consecutive series of values which were either all above or all below the median for the total sample (one-sample runs test, Siegel, 1956).

C. microlepis exhibited a highly developed system of Lateral Displays; in *Corynopoma*, aggressive interactions were more sporadic and not as stereotyped. In *C. microlepis*, dominance was stable and highly polarized; a dominant animal was rarely challenged. In *Corynopoma*, dominance was often unstable and unpredictably shifted, especially among females. It is postulated that in *Corynopoma* the loss of much of the stereotyped characid aggressive behavior has been accompanied by a loss of aggressive tendency and an instability of those dominance interactions that remain. This will be further discussed in a following section. Here it is sufficient to point out that as a probable result of these differences in individual recognition and degree of aggressive interactions, the time course of change in dominance and courtship relations is in *Corynopoma* at the level of the daily cycle, whereas in *C. microlepis* it may be on the order of months.

Changes in feeding behavior appeared in association with the length of time since previous feeding. In *Corynopoma* in particular, the degree of response to a cloud of *Drosophila* released over the surface was much greater after several days of starvation than when the fish had just finished a heavy meal of brine shrimp (although even then a certain amount of breaking the surface was seen). Again, frequency of Nipping Surface was in that species correlated with time since previous feeding. As mentioned previously, this supports the hypothesis that Nipping Surface is properly a feeding rather than "comfort" movement. But, as Nipping Surface by the male, but not by the female, was positively correlated with the number of courtship sequences (Nelson, in press), it is perhaps an act of compound motivation. Irrespective of whether fluctuations in the probability of Nipping Surface by the male *Corynopoma* can be analyzed into feeding and courtship components, it is important to point out that during the tropical dry season food is likely to be in short supply, and Nipping Surface is thus likely to be more frequent than at times of the year when food is more abundant. This should be kept in mind when consideration is given to the hypothesis on the evolution of rhythmic Gulping and Croaking in *Glandulocauda*, presented below.

In conclusion it should be emphasized that at these higher levels of the temporal hierarchy, no activity is seen to occupy a period of time exclusively, with the possible exception of spawning. Thus, within the course of a day a *Corynopoma* male may court, feed, fight, partake in repeated aggregation and dispersal, and change its dominance relations several times. This is to be contrasted, for example, with the behavior of a male Stickleback (Tinbergen, 1951), in which long periods of time may be occupied successively by nest-building, territorial defense, courtship, and finally parental care, each to the near exclusion of other activities. Thus, in the case of the Stickleback one may place a much greater degree of confidence in the reality and biological significance of these higher levels of the hierarchy of behavioral organization in time. In fact, it is suggested that whatever temporal hierarchy is found in the internally-fertilizing Glandulocaudines is, at least at these higher levels, largely imposed by the environment, and, at these higher levels, may not closely reflect any hierarchically organized physiological substructure. In this connection the temporal separation between mating and spawning which is made possible by internal fertilization should be remembered. In the Glandulocaudines, a new physiological background for courtship and mating in the female had to be developed, and there is little *a priori* reason to think that this background has to evolve from "reproductive drive." In the male, it was advantageous to have an opportunistic readiness to court at all times, but the potentialities for feeding, fighting, aggregation, etc., presumably had to be retained. Thus, behavior may be expected to be organized temporally and physiologically on a hierarchical basis only if it has been advantageous to have such an organization.

Short-term temporal patterns of behavior.—A full discussion of this topic as it relates to glandulocaudine courtship is given elsewhere (Nelson, in press). Here only those aspects of the subject which bear upon the evolution of behavior in this group will be considered.

Three general sorts of temporal pattern may be distinguished in glandulocaudine courtship. First, the general pattern of male courtship in *Corynopoma, Coelurichthys,* and *Glandulocauda* sequences was characterized by relatively stationary probabilities of occurrence of the various acts, and there was little evidence that the performance of one act affected the probability of performance of other acts very far removed from it in time. In particular, there was no evidence that the putative act of sperm transfer, Rolling, exerted a depressive effect upon further courtship. If one considers the lowering of the probability of occurrence of acts belonging to the same broad class (in this case, courtship acts) as the criterion, Rolling cannot be regarded to be consummatory in these genera. Similar arguments apply to attempts to delineate appetitive behavior in courtship of the internally fertilizing Glandulocaudines. In fact, the concepts of appetitive and consummatory behavior appear to be irrelevant to a discussion of courtship in these species. Hinde (1958), who has devoted considerable attention to this problem, came to the similar conclusion that the appetitive-consummatory dichotomy did not contribute to an understanding of the changes underlying the temporal pattern of canary nest-building behavior.

A second broad class of pattern seen in glandulocaudine behavior may be illustrated by the non-stationary nature of the courtship sequence in *Pseudocorynopoma.* Here, for example, the probability of Wobbling by the male rose to a peak, abruptly fell, and slowly rose again. Wobbling may be considered consummatory in that its performance was accompanied by a marked depression in the probability of its own recurrence; however, there is little direct evidence that this effect was a result of the performance of Wobbling itself rather than of a falling level of some internal process of which Wobbling was itself only a reflection. There is reason to expect, however, that the rising phase of the Wobbling cycle was a result of the cumulative, excitatory effect of the performance of courtship itself upon the probability of occurrence of further courtship. In *Corynopoma,* the occurrence of each courtship act appeared to exert a non-cumulative effect upon the occurrence of the immediately following event (Nelson, in press). It appears likely that *Pseudocorynopoma doriae,* an apparently externally fertilizing species with such features associated with internal fertilization as the caudal gland, is descended from an internally fertilizing Glandulocaudine. In this case all that is necessary to convert the pattern of such a form as *Corynopoma* into that of *Pseudocorynopoma* is a lengthening of the period over which the influence of the performance of one act extends.

Similar cumulative effects were seen in the other Glandulocaudines. In *Coelurichthys*, male behavior during the course of an encounter apparently acted cumulatively to lower the tendency of the female to flee, which in turn brought about a lowering in the probability of occurrence of Chasing by the male, and a resulting increase in courtship sequence length. Similarly, in *Corynopoma* there was evidence that male Displays acted cumulatively in raising the probability of female responses (Nelson, in press).

Still a third sort of temporal pattern is seen in the rhythmic repetition of Gulping and Croaking in *Glandulocauda*. Here, Gulping could be considered to be consummatory in that it lowered the probability of its own recurrence, but little seems to be gained by this. In *Corynopoma* and other species, Nipping Surface occurred not rhythmically but randomly in time (Nelson, 1963). There is every reason to believe that Gulping in *Glandulocauda* is derived from Nipping Surface. If this is so, then the rhythmicity of Gulping forms a very interesting case of the transformation of a temporal pattern to suit new needs. Elsewhere (Nelson, 1963), it is postulated that the lower limit of the inter-Gulp interval is imposed by the fact that at each Gulp the Hovering, Croaking male's attention is momentarily distracted from the female, so that it may be disadvantageous to Gulp more frequently than is necessary to sustain Croaking, and that the upper limit is set by the amount of air—utilized in producing Croaking—which may be taken up at one Gulp. Thus, it seems that the periodic, consummatory pattern of Gulping developed from the random pattern of Nipping Surface only because it was advantageous for Gulping to recur at regular intervals, and not because of any intrinsic tendency of animal's physiological processes to be organized in a consummatory fashion.

THE EVOLUTION OF BEHAVIORAL ACTS AND THEIR PATTERNS IN THE GLANDULOCAUDINES

With the possible exceptions of Tail-Beating and Head-to-Head Lateral Displays in *Coelurichthys* and *Glandulocauda*, and of Upright Postures in *Pseudocorynopoma*, aggressive displays in these three genera are quite similar to those in the generalized characids studied. They are in both groups highly correlated in time with Chasing and Biting. An intermediate between the Lateral Displays and the Upright Posture of *Pseudocorynopoma* was seen in the tilting upward of the heads during Double Broadsides in *C. tenuis*.

In *Corynopoma*, only variants of the Upright Posture appear to remain as aggressive displays. Even these were generally only seen in response to an Approaching animal, and were more-or-less confined to male-male encounters. Accompanying this loss of stereotyped aggressive interactions has been a reduction in the amount of Chase-Biting seen, and a weakening of temporal correlation between Biting and Upright Postures. *Corynopoma* appears to be among the least aggressive of characids.

Two factors, internal fertilization and the surface-feeding habit, appear to have been of overwhelming importance in the evolution of courtship acts and their temporal pattern in the Glandulocaudines. First, it would appear that the habit of taking surface food was in some way incorporated into mating above the surface of the water. In *Corynopoma*, frequency of Nipping Surface in the male was significantly correlated with the number of courtship sequences (Nelson, in press) and it is possible that a similar correlation facilitated its incorporation into mating above the surface. In the Characidae, this manner of mating has developed independently three times, all in surface-feeding forms. It was earlier argued that the habit of mating above the surface in its turn facilitated the development of internal fertilization itself, and that with the temporal separation of mating and spawning, the predominant factor governing the evolution of courtship behavior and elaborate dimorphic structures in the Glandulocaudini was the need to ensure successful fertilization of an unresponsive female.

The caudal gland and behavior associated with it may be discussed first. A superficial inspection of courtship in *Corynopoma, Pseudocorynopoma, Glandulocauda,* or *C. microlepis* reveals no striking features that might be correlated with caudal gland morphology. However, in *C. tenuis*, Dusting (fig. 8, c) appears to be ideally suited to wafting an odorous substance from the caudal gland to the region of the female's head, and inspection of the caudal morphology in this species reveals that it too would be fitted for this purpose. Two fin rays are modified to form opposing halves of a tube, and are so arranged that the edge of one fits inside the other (fig. 3, f), so that any differential movements of the two fin rays will cause them to operate as a bellows. It is postulated that the caudal gland produces a substance which, when directed toward the female by this bellows during Dusting, increases the probability that she will Pair.

Reinspection of the caudal glands of the other Glandulocaudines indicates that in all cases, the accessory tube, pouch, or trough is such that movement of the caudal fin will result in water being sucked in and expelled over a glandular surface, (figs. 3 and 4). A possible exception is *Glandulocauda* (fig. 5, a and b), in which modified scales and fin rays are present, but a scaly pouch and a fin-ray tube are lacking. In *Glandulocauda* it may be that Croaking has taken over much of the function of the caudal gland. This suggests that the behavior of the other species be re-examined to see if a particular movement appears to be associated with the caudal gland. In *Corynopoma* the most likely candidates are the peculiar figure-of-eight Chases in front of the female. In *Pseudocorynopoma*, Ovalling in particular and also Zigzagging could result in scent-laden water being directed toward the female, and in *C. microlepis* the high frequency, encircling Zigzagging might have this function. A possible scent-directing movement appears to be absent in *Glandulocauda*, which as mentioned appears to Croak instead. Thus, the correlation seems complete, and it may be said that the most likely function of the caudal gland is to produce a substance which, when wafted toward the female by accessory structures and specialized courtship acts, increases the probability of successful mating.

The resemblance between the figure-of-eight Chasing in *Corynopoma* and Dusting in *C. tenuis* was not seen until the analysis of *Corynopoma* courtship was completed. It is possible that they have evolved from some common ancestral act, performed in front of the female. As *Corynopoma* has evidently evolved from a form similar in general and caudal morphology to *Gephyrocharax* (see fig. 4, *e* and *f*), and as *Gephyrocharax* appears biogeographically to be generalized cosmopolitan and is fairly close to the morphologically generalized genera *Phenacobrycon* and *Argopleura,* its study should be most rewarding.

In any event, the position of the male *Corynopoma* relative to the female in the paddle displays is quite similar to that of the male *C. tenuis* during Dusting, and the timing and role of female Following appear similar in both species. Eigenmann (Eigenmann and Myers, 1929) has commented upon the "development of movable spots near the middle of the body" in non-homologous structures in the Glandulocaudines: the tip of *Corynopoma*'s paddle, a paddle-like scale extension in *Pterobrycon*, and the pectorals tipped with black in *Gephyrocharax melanocheir* and *Ps. doriae* (not seen in the present material). It is at least likely that these all have similar functions. The evolution of the opercular extension in *Corynopoma* is envisioned as follows. During Yawning in all Glandulocaudines the opercles and median fins are raised (fig. 5, *c*). I postulate that in the ancestor to *Corynopoma,* median fin unfolding was part of the courtship pattern, and that this, whether or not it evolved from Yawning, was accompanied by mouth opening and gill-cover raising as is seen now in *Corynopoma* Twitching and Shaking (fig. 10, *b*). In another Glandulocaudine, *Diapoma* (Eigenmann and Myers, 1929 pl. 67, fig. 2), the opercle is notched above and prolonged below, and forms an intermediate condition between the slightly prolonged opercle of the female *Corynopoma* and the male paddle. It is possible that even this slight a prolongation in the ancestral *Corynopoma* was sufficient to elicit female Biting, and that this latter became incorporated into courtship as Nipping. The perfection of this Nip-eliciting device resulted in the elongate *Corynopoma* paddle. A similar but not as spectacular evolutionary sequence is postulated for the tipping of the pectorals with black in *G. melanocheir* and certain populations of *Ps. doriae;* it is tempting to speculate that these latter are the ones which have produced the reports of internal fertilization in *Pseudocorynopoma*. It is even more tempting to speculate that *Pterobrycon* arose from *Corynopoma* as a homeotic mutant (Goldschmidt, 1938). In fishes, the scales are serial homologs of the dermal bones of the skull, including the suboperculum from which the opercular extension is developed (Romer, 1955).

What of the function of paddle-Nipping? It was observed that a properly oriented Quivering, as in figure 10, *c*, was rare unless preceded by Nipping. If Nipping did not precede Quivering, the male was generally positioned too far forward on the female, so that his median fins enveloped her head; she usually soon escaped and Pairing and Rolling were not seen. I postulate that the function of Nipping is to inform the male that the female is properly oriented for Quivering to begin; it has been demonstrated (Nelson, in press) that Nipping following Twitch raises the probability of occurrence of Quiver. It may be that in *Gephyrocharax melanocheir* and the *Ps. doriae* with the black-tipped pectorals, and in *Pt. landoni,* Nipping at the "movable spot" also occurs.

A properly oriented Quivering may aid in ensuring not only that Pairing and Rolling will occur, but also that they too will be properly oriented. In most of the generalized Characidae, hooks are present on the male's anal fin, and in many on the pelvic fins as well. Caudal hooks, on the other hand, are held by Böhlke (1958) to be unusual among the Characidae, but were found in five of the glandulocaudine genera studied, and elsewhere only in one specimen of *Compsura*. The function of pelvic and anal hooks in the generalized characids appears to be to maintain proximity of the quite active pair during spawning, but they may also be viewed as a preadaptation facilitating the development of internal fertilization, with which development the accurate orientation of a pair takes on a new significance, especially in the absence of an intromittent organ. This is viewed as the factor behind the development of hooks on the caudal, and in *Acrobrycon* even the dorsal fin; in *Coelurichthys* this is believed to have led to the single row of considerably enlarged and deepened hooks on the anterior fin rays. The absence of pelvic and caudal hooks in the present material of *Ps. doriae* thus becomes explicable, as the evidence indicates that in these internal fertilization has been lost. In any event, in view of the various hooks and encircling fins, it may be difficult for a *Corynopoma* pair to reorient with respect to one another once Quivering has begun.

Zigzagging in *C. microlepis* and *Pseudocorynopoma* appears similar, and may without difficulty be derived from the up-and-down Chasing in *Corynopoma*, via Zigzagging in *C. tenuis* or *G. inequalis*. In the latter species Zigzagging appears to be still an intention movement of rising to the surface, and it is likely that in the other groups it arose in a similar fashion. While in *Glandulocauda* it appears very definitely to be an intention movement of Gulping (or originally, Nipping Surface), it is not known whether in the other species it arose directly from Nipping Surface, or indirectly as an intention movement of pairing above the surface. The lateral undulations of the upward movement in *Pseudocorynopoma* and *C. microlepis* suggest the latter origin; the lack of pronounced lateral undulations in *Corynopoma* and *C. tenuis* suggests rather a direct derivation from intention movements of Nipping Surface.

Zigzagging in *Pseudocorynopoma* often merged imperceptibly into Wobbling as the upward component became more and more horizontal, and in the nature of the lateral undulations they were quite similar. The female was often seen to respond to Wobbling by Wobbling also; when this occurred Impinging nearly always followed. The parallels with *Corynopoma* Quivering followed by Pairing (= mutual Quivering) were striking; indeed the only things lacking were bodily contact and the rise to the surface. It is difficult to believe that they, and indeed the lateral undulations of Zigzagging and Pairing in the other species, are not derived from a common origin.

On this hypothesis, Impinging is at least temporally analogous to Rolling, but its derivation remains obscure. In most closely resembled *Pseudocorynopoma*'s aggressive Upright Posture, however, and the possibility of their mutual derivation from some common antecedent movement must be considered. The origin of courtship displays from aggressive activities appears to be fairly common (Tinbergen, 1952) but the present author knows of no case in which the actual mating or spawning act is derived from an aggressive movement. It is more likely, therefore, that both the Upright Posture and Impinging are derived from some prior activity, possibly originating in courtship.

Chasing in *Corynopoma* and Approaching in *Pseudocorynopoma* are orientation movements, and offer no difficulties of interpretation. Chasing in *Glandulocauda* and *Coelurichthys* may be regarded as continuous orientation movements to a rapidly moving female. Hovering in *Glandulocauda* and *Coelurichthys* is somewhat more difficult to interpret. It appears likely that it is also derived from arrested orientating movements; in *Glandulocauda* it possibly functions to maintain a proper orientation during Croaking. Perhaps in *Coelurichthys,* Hovering accustoms the female to the male and assures her of his nonaggressive intentions. Zigzagging and Dusting (and the paddle displays in *Corynopoma*) appear to require a relatively quiescent female for the maintenance of the proper orientation. In *Coelurichthys* and *Glandulocauda* a female was never observed to escape from a Hovering male without a Chase beginning, but often, female *Coelurichthys* were "misplaced" by Zigzagging or Dusting males.

Ovalling in *Pseudocorynopoma* cannot be derived with ease from any other observed glandulocaudine movement. Although it bore resemblances to both the alternating orientation of Hovering in *Glandulocauda* and the figure-of-eight Chasing of *Corynopoma,* homologizing Ovalling with these appears to be unsafe. However, as mentioned above, Ovalling and figure-of-eight Chasing may both serve in the dissemination of the caudal gland substance.

In the temporal patterning of courtship sequences, *Coelurichthys, Glandulocauda,* and *Corynopoma* appear to be closely similar, but each has its special modifications which for the most part appear to be associated with its needs. Thus, *Glandulocauda* has added to a more-or-less random sequence the regular repetition of Gulping and Croaking. *Pseudocorynopoma* courtship patterning, although superficially quite distinctive, also has resemblances to the temporal patterning found in the other species (Nelson, in press). Again, the distinctiveness of the *Pseudocorynopoma* temporal pattern may be adapted to its needs. Commonly among such generalized characids as *Hyphessobrycon* (pers. obs.), there will be many successive spawning acts during each of which only a few eggs are deposited; these will be separated by intervals of a few to fifteen minutes. In the internally fertilizing species, eggs are similarly deposited singly or a few at a time, at intervals of several minutes. It appears likely that, at least in terms of temporal patterning of spawning, *Ps. doriae* has reverted to the generalized characid condition. It is possible that some feature of characid physiology permits only a few eggs to be deposited at a time; in this event, there is no need for the male continuously to engage in intense courtship, and indeed it may be that periodic rests are required of the male as well. Thus it appears that the courtship pattern of the male *Pseudocorynopoma* may be tuned to the spawning patterns of the female.

It may be mentioned again that in an internally fertilizing tropical species it may be advantageous to maintain courtship readiness in the male over much or even all of the year, yet this must not be allowed to interfere seriously with his other biological needs. A way to accomplish this is to do away with drastic fluctuations in internal motivation, and to substitute a close dependence upon the environmental stimulus of presence of a nearby female for the initiation of courtship. In *Coelurichthys* and *Glandulocauda* there is a high degree of aggressiveness in both sexes, for whatever reasons, and this must be overcome before courtship can be successful. This, it is argued, has resulted in the non-stationary probabilities between courtship sequences, almost as a by-product. Otherwise, in *Glandulocauda* and *Coelurichthys,* as well as in *Corynopoma,* the stationary and relatively indeterminate nature of courtship appears to be a good way of maintaining a relatively high, relatively constant probability that the male will court when he gets the chance, and will be successful at it. Again, the temporal pattern seems adapted to the needs.

The live-bearing poeciliid *Lebistes reticulatus* appears to have a much more elaborate courtship pattern than do the Glandulocaudines, but one gathers from the discussion of Baerends *et al.* (1955) that its pattern too may be very loosely organized, with relatively stationary probabilities of occurrence for different activities. Here, and in other internally fertilizing groups such as the Eider ducks (McKinney, 1961), the female remains relatively passive throughout the precopulatory sequence, and the male courtship pattern seems to be rather loose. This appears to be characteristic of many internally fertilizing species, and in fishes is to be contrasted with courtship in the Cichlidae (Baerends and Baerends-van Roon, 1950) and Gasterosteidae (for example, Morris, 1958). It may be that the passive glandulocaudine female supplies the male with relatively few behavioral triggers upon which he can rely for the ordering of his behavioral pattern; forced to depend almost entirely upon feedback stimuli from his own performance, he produces a relatively indeterminate sequence of acts (Nelson, in press).

In conclusion, the sheer number of events in glandulocaudine courtship which can be traced directly or indirectly to Nipping Surface is striking. Pairing and Rolling, Zigzagging, up-and-down Chasing, Gulping, and Croaking, all appear to owe a debt to the glandulocaudine habit of feeding from the surface. Surface-feeding appears to have greatly facilitated the development of internal fertilization; the temporal pattern of courtship, the caudal gland and its accessory structures, all the other striking morphological features of the male, as well as practically all of the courtship acts that are not directly attributable to Nipping Surface (and indeed even most of those that are), all seem to have evolved in the service of this major reproductive adaptation.

SUMMARY AND CONCLUSIONS

1. The Glandulocaudini are internally fertilizing, surface feeding, South American characid fishes. The present paper gives the results of an investigation of their behavioral and morphological adaptations. The behavior of six species, belonging to four genera, was studied in aquaria. Field observations were made of one species, *Corynopoma riisei,* in Trinidad. In addition, preserved material of most of the glandulocaudine genera and of certain related groups was examined.

2. Locomotion in the Glandulocaudines was found to be similar to that in the generalized characids. All species appeared adapted for surface-feeding, and often nipped at the water surface even when food was not present. Spawning occurred in the absence of males. Eggs were placed rather than scattered by the females, and there were no indications of care of eggs or young. The putatively externally fertilizing Glandulocaudine, *Pseudocorynopoma,* did not spawn.

3. Aggregation was seen in all species, often in the absence of evident external

stimuli. Aggregation in relation to other behavior is discussed. The development of aggregation and other behavior in three species is outlined and discussed.

4. Interspecies differences in aggressive behavior are described. In *Corynopoma*, aggressive displays were weakly developed, sporadic in occurrence and not well correlated with Biting. Dominance hierarchies could, however, be formed. In *Coelurichthys, Pseudocorynopoma,* and *Glandulocauda,* aggressive displays were strongly developed in both sexes and closely correlated with Biting.

5. Individual recognition was marked in *Coelurichthys microlepis'* aggressive and courtship encounters, more weakly developed in *Corynopoma,* and least noticeable in *Coelurichthys tenuis.* Correlations of individual recognition with aggressive behavior, the courtship pattern, and ecological differences are discussed, and the individual differences which may make it possible are examined.

6. In *Corynopoma,* the immediately preceding experience of the males was found to play a large role in determining the outcome of an encounter. This was less marked in *Coelurichthys.* The role of this "set" is discussed in relation to individual recognition and to the nature of the courtship pattern in this species.

7. Ethograms are given for social activities. Many courtship acts in all species examined showed direct or indirect relations to the act of feeding at the surface, and to the act of mating, which in the internally fertilizing species occurred above the surface of the water. The absence of pairing above the water in *Pseudocorynopoma* could be attributed to the recorded absence of internal fertilization in that genus.

8. Courtship in *Corynopoma* is described. Sequences of courtship could be defined statistically. Within and between sequences the various measures of male courtship remained relatively stationary. The act of fertilization did not appear to be consummatory. Sequences of courtship were accompanied by a darkening of the male belly spot. The rare female responses were necessary for the occurrence of the series of acts leading to fertilization.

9. During a courtship sequence, in *Ps. doriae,* the male's courtship activity rose to a peak and then abruptly terminated, to slowly rise again, in the fashion of a relaxation oscillation. Although direct evidence was not available, it was concluded that the pattern was most likely a result of cumulative facilitatory effects of male courtship upon his own performance. Female responses were found to alter the male behavior pattern in the direction of the putative spawning act. This latter, Impinging, occurred well beneath the surface of the water, and did not resemble Rolling, the act of fertilization in the other glandulocaudine genera.

10. In *Coelurichthys* and *Glandulocauda,* a courtship encounter consisted of alternating aggressive and courtship sequences, with the former gradually becoming shorter and less frequent. In *C. tenuis,* analysis indicated that the male pattern within a courtship sequence was rather indeterminate and stationary. The probabilities of occurrence of various acts were not stationary from sequence to sequence; rather, the probability of Chasing fell during an encounter in a stepwise fashion, apparently as a result of the cumulative effect of male activity upon the female.

11. In *Glandulocauda,* a pattern of rhythmic Gulping of air by the male was superimposed upon a courtship sequence which was in general quite similar to that

in *Coelurichthys;* Gulping was found to be highly correlated with a sound, Croaking, produced by the courting male as he Hovered above or below the female.

12. Study of preserved material showed that the unifying glandulocaudine features were the presence in the male of glandular areas at the base of the tail and the features associated with surface life. On the basis of this and other evidence it was concluded that the group was monophyletic in origin and that its basic adaptations were to surface feeding and internal fertilization.

13. The evolution of internal fertilization appears to be a response to a tropical habitat with well-marked wet and dry seasons. In this situation the temporal separation of mating and spawning may be advantageous. Top-feeding appears to have facilitated the development of this temporal separation *via* internal fertilization. As a result, the males of the internally fertilizing species show no strong seasonality of courtship, but the female remains tied to the rainy reason for spawning. This in turn has governed the evolution of the physiological bases and temporal patterning of courtship in this group. In the internally fertilizing Glandulocaudines a continuously active male courts an unresponsive female; his efforts raise in a cumulative fashion the probability of successful fertilization.

14. To this end have been developed the sundry features of glandulocaudine sexual dimorphism: the caudal glandular areas, various accessory tubes, pouches and furrows which appear to direct a pheromone toward the female during courtship, courtship acts facilitating the transmission of this substance, greatly enlarged fins in the males of several species, the development of a greatly prolonged opercular extension in *Corynopoma* and specialized behavior associated with it, Croaking in *Glandulocauda* males, and the development of dorsal and caudal hooks.

15. It is made clear that the particular temporal pattern of behavior which is found, and its physiological bases, will be adaptive, but will bear a relation to the evolutionary history of the species. Thus, in the Glandulocaudines the absence of appetitive and consummatory phases of male courtship may be attributed to the advantage the male gains in being continuously sexually active. In *Pseudocorynopoma* there has been a return to a courtship pattern the performance of which lowers the probability of its own recurrence, but this pattern has not arisen *de novo*, but rather shows traces of the temporal organization in the internally fertilizing species. In all species the effect of male activity upon the female appears to be cumulative. This dependence upon male activity may be a substitute for the dependence of female courtship upon the physiological events underlying egg maturation, which has been lost in the course of evolution of a temporal separation between mating and spawning.

LITERATURE CITED

ALLEE, W. C.
    1951. Cooperation among Animals, with Human Implications. New York: Abelard-Schuman Ltd.

ALTMAN, S. A.
    1962. A field study of the sociobiology of rhesus monkeys, *Macaca mulatta*. Ann. N. Y. Acad. Sci. 102:338–435.

BAERENDS, G. P.
    1958. The contribution of ethology to the study of the causation of behaviour. Acta Physiol. Pharmacol. Neerl. 7:466–499.

BAERENDS, G. P., and J. M. BAERENDS–VAN ROON
    1950. An introduction to the study of the ethology of cichlid fishes. Behaviour, supp. I:1–242.

BAERENDS, G. P., R. BROUWER, and H. T. J. WATERBOLK
    1955. Ethological studies on *Lebistes reticulatus* (Peters). I. An analysis of the male courtship pattern. Behaviour VIII: 249–334.

BARNARD, K. H.
    1941. Note on alleged viviparity in *Barbus viviparus* and description of a new species of *Beirabarbus*. Ann. Mag. Nat. Hist. (11) 8:469–471.

BEACH, F. A.
    1954. Problems in scoring and terminology. Ms., Genetic, psychological, and hormonal factors in the establishment and maintenance of patterns of sexual behavior in mammals. 99–124.

BEEBE, W.
    1945. Vertebrate fauna of a tropical dry season mud-hole. Zoologica (N. Y.) 30:81–88.

BERG, L. S.
    1940. [Classification of fishes both recent and fossil.] Doklady Zool. Inst. Akad. Nauk SSSR, 5(2):87–517. Russian with English translation; photo lithoprint, 1947, by J. W. Edwards. Ann Arbor, Michigan.

BERTIN, L.
    1958. Viviparité des téléostéens. *In* P.-P. Grassé (ed.), Traité de Zoologie XIII: 1791–1812.

BOESEMAN, M.
    1960. The freshwater fishes of the island of Trinidad. *In* P. W. Hummelinck (ed.): Studies on the fauna of Curacao and other Carribbean islands. X:72–153.

BÖHLKE, J.
    1954. Studies on fishes of the family Characidae, 7. A new genus and species of glandulocaudine characids from central Brazil. Stanford Ichth. Bull. 4:265–274.
    1958a. Studies on fishes of the family Characidae, 14. A report on several extensive recent collections from Ecuador. Proc. Acad. Nat. Sci. Phila. 110:1–122.
    1958b. Results of the Catherwood Foundation Peruvian Amazon Expedition. The description of two new xenurobryconine characids. Copeia 1958:318–325.

BREDER, C. M.
    1926. The locomotion of fishes. Zoologica (N. Y.) 4:159–297.
    1951. Studies on the structure of the fish school. Bull. Am. Mus. Nat. Hist. 98:1–28.

BREIDER, I.
    1948. Weitere Untersuchungen zur Fortpflanzungsbiologie der Characidae. Wchnschr. Aquar. u Terrarienk. 42:9–12.

BROOKS, J. L.
    1950. Speciation in ancient lakes. Quart. Rev. Biol. 25:30–60, 131–176.

CLARK, E., and L. R. ARONSON
    1951. Sexual behavior in the Guppy, *Lebistes reticulatus* (Peters). Zoologica (N. Y.) 36: 49–66.

CLOTHIER, C. R.
    1950. A key to some Southern California fishes based on vertebral characters. Calif. Fish and Game Bull. 79.

DICE, L. R.
  1952. Natural Communities. Ann Arbor: Univ. Mich. Press.
DUMPERT, V.
  1921. Zur Kenntnis des Wesens und der physiologischen Bedeutung des Gahnens. J. f. Psychol. u. Neurol. 27:82–95.
EIGENMANN, C. H.
  1911. New Characids in the collection of the Carnegie Museum. Ann. Carnegie Mus. 8:164–181.
  1914. On new species of fishes from the Rio Meta Basin of Eastern Colombia, etc. Indiana Univ. Studies 23:229–240.
  1915. The Cheirodontinae, a subfamily of minute Characid fishes of South America. Mem. Carnegie Mus. 7:1–100.
  1917. The American Characidae, part I. Mem. Mus. Comp. Zool. Harvard XLIII 1:1–102.
  1927. The American Characidae, part IV. Mem. Mus. Comp. Zool. Harvard XLIII 4:318–428.
EIGENMANN, C. H., and G. S. MYERS
  1929. The American Characidae, part V. Mem. Mus. Comp. Zool. Harvard XLIII 5:429–558.
EVERMANN, B. W., and E. L. GOLDSBOROUGH
  1909. Notes on some fishes from the Canal Zone. Proc. Biol. Soc. Wash. XXII:95–104.
FRASER-BRUNNER, A.
  1948. A spoon-fed female. Aquarist (London) 13:162.
GOLDSCHMIDT, R.
  1938. Physiological Genetics. New York: McGraw-Hill.
GREGORY, W. K., and G. M. CONRAD
  1938. Phylogeny of the characid fishes. Zoologica (N. Y.) 23:319–360.
GUPPY, P. L.
  1934. Observations on Trinidad larvicidal fishes. Trop. Agri. 11:117–122.
HESSE, R., W. C. ALLEE, and K. P. SCHMIDT
  1951. Ecological Animal Geography. 2nd ed., New York: Wiley.
HINDE, R. A.
  1958. The nest-building behavior of domesticated canaries. Proc. Zool. Soc. Lond. 131:1–48.
HOEDEMAN, J. J., and J. C. M. DE JONG
  1949. Encyclopedia of Water Life. Amsterdam: "De Regenboog."
INNES, W. T.
  1948. Exotic Aquarium Fishes. 9th ed., Philadelphia: Innes.
KORTLANDT, A.
  1940. I. Eine Übersicht der angeborenen Verhaltensweisen des Mittel-Europäischen Kormorans (*Phalacrocorax carbo sinensis* Shaw and Nodd.), ihre Funktion, ontogenetische Entwicklung, und phylogenetische Herkunft. II. Wechselwirkung zwischen Instinkten. Arch. neerl. Zool. 4:401–520.
KULKARNI, C. V.
  1940. On the systematic position, structural modifications, bionomics and development of a remarkable new family of cyprinodont fishes from the province of Bombay. Rec. Indian Mus. XLII (2):379–423.
KUTAYGIL, N.
  1958. Insemination, sexual differentiation, and secondary sex characters in *Stevardia albipinnis* Gill. Istanbul Univ. Fen Fakultesi Mecmuasi (B) XXIV:93–128.
MCKINNEY, F.
  1961. An analysis of the displays of the European Eider *Somateria mollissima mollissima* (Linnaeus) and the Pacific Eider *Somateria mollissima* var. *nigra* Bonaparte. Behaviour, supp. VII.
MEES, G. F.
  1961. Description of a new fish of the family Galaxiidae from Western Australia. J. Roy. Soc. Western Australia 44:33–38.

MIRANDA-RIBEIRO, A. de
- 1908. Peixes da Ribeira. Kosmos (Rio de Janeiro) V (2).
- 1939. Alguns novos dados ictiologicos da nossa fauna. Bol. Biol. (n. s.) 4:358–363.

MORRIS, D.
- 1958. The reproductive behaviour of the Ten-spined Stickleback (*Pygosteus pungitius* L.). Behaviour, supp. VI:1–154.

MYERS. G. S.
- 1958. Trends in the evolution of Teleostean fishes. Stanford Ichth. Bull. 7:27–30.

MYERS, G. S., and J. BÖHLKE
- 1956. The Xenurobryconini, a group of minute South American Characid fishes with teeth outside of the mouth. Stanford Ichth. Bull. 7:6–12.

NELSON, K.
- 1963. Behavior and morphology of the glandulocaudine fishes (Ostariophysi, Characidae). Doctoral thesis, Univ. of Calif., Berkeley. Univ. microfilms.
- in press. Temporal patterning of courtship behaviour in the glandulocaudine fishes. Behaviour XXIV (1–2).

NICHOLS, J. T.
- 1913. On two new Characins in the American Museum. Proc. Biol. Soc. Wash. 26:151–152.

NIKOLSKII, G. V.
- 1961. Special Ichthyology. 2nd ed. [Translated from Russian.] Jerusalem: Israel Program for Scientific Translations, Ltd.

PICKFORD, G. E., and J. W. ATZ
- 1957. The Physiology of the Pituitary Gland in Fishes. New York: New York Zoological Soc.

PRICE, J. L.
- 1955. Survey of the freshwater fishes of the island of Trinidad. J. Agr. Soc. Trin. Tobago 55:390–416.

RAMIREZ E., M. V.
- 1960. Nuevos peces para la fauna Venezolana. Mem. Soc. De Ciencias Naturales "La Salle" 56 (May.–Aug.).

REGAN, C. T.
- 1907. Descriptions of two new Characinid fishes from South America. Ann. Mag. Nat. Hist. (7) 20:402–403.

ROMER, A. S.
- 1955. The Vertebrate Body. 2nd ed., Philadelphia: Saunders.

ROSEN, D. E., and A. TUCKER
- 1961. Evolution of secondary sex characters and sexual behavior patterns in a family of viviparous fishes (Cyprinodontiformas : Poeciliidae). Copeia 1961:201–212.

SCHULTZ, A. V.
- 1962. The Dragonfin. Trop. Fish Hobbyist X (June):5–9.

SCHULTZ, H.
- 1959. The *Mimagoniates* species. Trop. Fish Hobbyist VII (June):46–53.

SCHULTZ, L. P.
- 1959. The generic status of *Mimagoniates* and *Glandulocauda,* South American characid fishes. Trop. Fish Hobbyist VIII (Oct.):6–11, 63–64.

SCHWAB, E.
- 1939. Der Befruchtungsvorgang und andere Beobachtungen bei *Glandulocauda inequalis.* Wchnschr. Aquar. Terrarienk. 36:17–19.

SIEGEL, S.
- 1956. Non-parametric Statistics for the Behavioral Sciences. New York: McGraw-Hill.

STEINDACHNER, E.
- 1878. Die Süsswasserfische des südöstlichen Braziliens, III. Sitzberichte Akad. Wien 74:559–664.

STERBA, G.
   1959. Süsswasserfische aus Aller Welt. Liepzig/Jena: Urania-Verlag.

TAVOLGA, W. N.
   1956. Visual, chemical and sound stimuli as cues in the sex discriminatory behavior of the gobiid fish, *Bathygobius soporator*. Zoologica (N. Y.) 41:49–64.

TINBERGEN, N.
   1951. The Study of Instinct. Oxford: Oxford Univ. Press.
   1952. 'Derived' activities; their causation, biological significance, origin, and emancipation during evolution. Quart. Rev. Biol. 27:1–32.

WEBER, M., and L. F. DE BEAUFORT
   1922. Fishes of the Indo Australian Archipelago, IV. Leiden: E. J. Brill.

WEITZMAN, S.
   1960a. Further notes on the relationships and classification of the South American characid fishes of the subfamily Gasteropelecinae. Stanford Ichth. Bull. 7:217–239.
   1960b. A contribution to the morphology and the classification of the fishes of the family Characidae. Doctoral thesis, Stanford Univ., Univ. microfilm Mic 60–3842.
   1962. The osteology of *Brycon meeki*, a generalized characid fish, with an osteological definition of the family. Stanford Ichth. Bull. 8:1–77.

WICKLER, W.
   1960. Die Stammegeschichte typischer Bewegungsformen der Fischbrustflosse. Z. f. Tierpsychol. 17:31–66.